STUDENT SOLUTIONS MANUAL

to accompany

APPLIED CALCULUS

SECOND EDITION

Deborah Hughes-Hallett
University of Arizona

Andrew M. Gleason
Harvard University

Patti Frazer Lock
St. Lawrence University

Daniel E. Flath
University of South Alabama

et al.

Prepared by

Elliot J. Marks
Harvard University

Ernie S. Solheid
California State University, Fullerton

Ray Cannon
Baylor University

JOHN WILEY & SONS, INC.

COVER PHOTO © Lester Lefkowitz/Corbis Stock Market.

To order books or for customer service call 1-800-CALL-WILEY (225-5945).

ISBN 0-471-21362-4

Printed in the United States of America

10 9 8 7 6 5 4 3 2 1

Printed and bound by Bradford & Bigelow, Inc.

CONTENTS

CHAPTER ONE

Solutions for Section 1.1

1. $f(35)$ means the value of P corresponding to $t = 35$. Since t represents the number of years since 1950, we see that $f(35)$ means the population of the city in 1985. So, in 1985, the city's population was 12 million.

5. **(a)** $f(30) = 10$ means that the value of f at $t = 30$ was 10. In other words, the temperature at time $t = 30$ minutes was 10°C. So, 30 minutes after the object was placed outside, it had cooled to 10 °C.

(b) The intercept a measures the value of $f(t)$ when $t = 0$. In other words, when the object was initially put outside, it had a temperature of a°C. The intercept b measures the value of t when $f(t) = 0$. In other words, at time b the object's temperature is 0 °C.

9. One possible graph is shown in Figure 1.1. Since the patient has none of the drug in the body before the injection, the vertical intercept is 0. The peak concentration is labeled on the concentration (vertical) axis and the time until peak concentration is labeled on the time (horizontal) axis. See Figure 1.1.

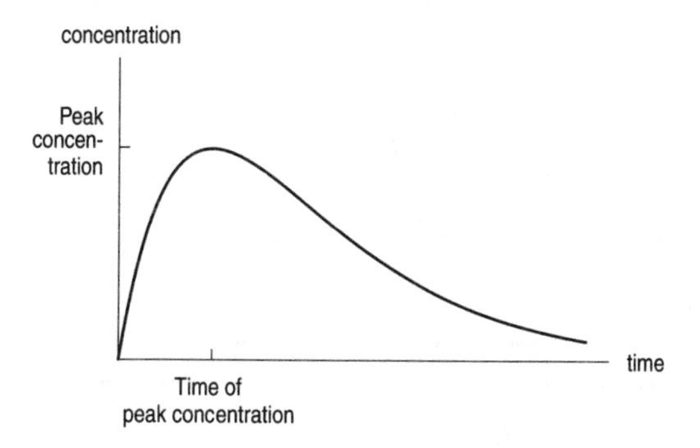

Figure 1.1

13. Substituting $x = 5$ into $f(x) = 2x + 3$ gives

$$f(5) = 2(5) + 3 = 10 + 3 = 13.$$

17. In the table, we must find the value of $f(x)$ when $x = 5$. Looking at the table, we see that when $x = 5$ we have

$$f(5) = 4.1$$

21. Figure 1.2 shows one possible graph of gas mileage increasing to a high at a speed of 45 mph and then decreasing again. Other graphs are also possible.

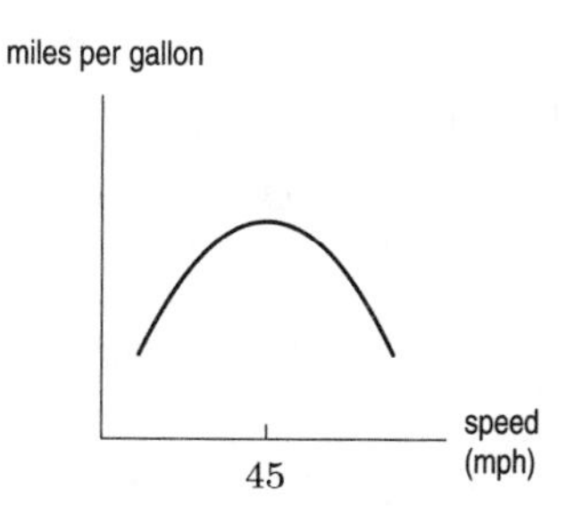

Figure 1.2

25. **(a)** We are asked for the value of y when x is zero. That is, we are asked for $f(0)$. Plugging in we get

$$f(0) = (0)^2 + 2 = 0 + 2 = 2.$$

(b) Substituting we get

$$f(3) = (3)^2 + 2 = 9 + 2 = 11.$$

(c) Asking what values of x give a y-value of 11 is the same as solving

$$\begin{aligned} y = 11 &= x^2 + 2 \\ x^2 &= 9 \\ x &= \pm\sqrt{9} = \pm 3. \end{aligned}$$

We can also solve this problem graphically. Looking at Figure 1.3, we see that the graph of $f(x)$ intersects the line $y = 11$ at $x = 3$ and $x = -3$. Thus, when x equals 3 or x equals -3 we have $f(x) = 11$.

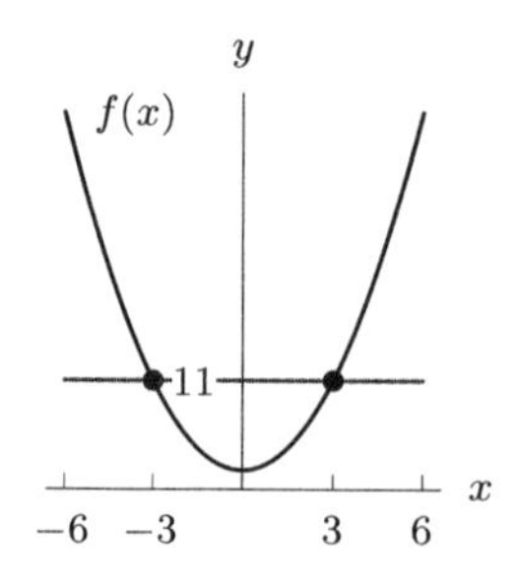

Figure 1.3

(d) No. No matter what, x^2 is greater than or equal to 0, so $y = x^2 + 2$ is greater than or equal to 2.

Solutions for Section 1.2

1. Rewriting the equation as

$$y = 4 - \frac{3}{2}x$$

shows that the line has slope $-3/2$ and vertical intercept 4.

5. The slope is $(1 - 0)/(1 - 0) = 1$. So the equation of the line is $y = x$.

9. **(a)** is (V), because slope is positive, vertical intercept is negative
(b) is (IV), because slope is negative, vertical intercept is positive
(c) is (I), because slope is 0, vertical intercept is positive
(d) is (VI), because slope and vertical intercept are both negative
(e) is (II), because slope and vertical intercept are both positive
(f) is (III), because slope is positive, vertical intercept is 0

13. **(a)** The slope is constant at 850 people per year so this is a linear function. The vertical intercept (at $t = 0$) is 30,700. We have

$$P = 30{,}700 + 850t.$$

(b) The population when $t = 10$ is predicted to be

$$P = 30{,}700 + 850 \cdot 10 = 39{,}200.$$

(c) We use $P = 45{,}000$ and solve for t:

$$\begin{aligned} P &= 30{,}700 + 850t \\ 45{,}000 &= 30{,}700 + 850t \\ 14{,}300 &= 850t \\ t &= 16.82. \end{aligned}$$

The population will reach 45,000 people in about 16.82 years, or during the year 2016.

17. Given that the equation is linear, choose any two points, e.g. (5.2, 27.8) and (5.3, 29.2). Then

$$\text{Slope} = \frac{29.2 - 27.8}{5.3 - 5.2} = \frac{1.4}{0.1} = 14$$

Using the point-slope formula, with the point $(5.2, 27.8)$, we get the equation

$$y - 27.8 = 14(x - 5.2)$$

which is equivalent to

$$y = 14x - 45.$$

21. We know that the linear approximation of this function must take on the form

$$Q = mt + b$$

where t is measured in years since 1986. The slope for the function would be

$$\text{slope} = \frac{\Delta Q}{\Delta t} = -1.24.$$

We also know that the function takes on a value of 19.72 at the year 1986 (i.e., at $t = 0$). Thus, if our function is to give the amount of gold in the reserve in years after 1986, it will take on the form

$$Q = -1.24t + 19.72.$$

25. **(a)** We are given two points: $S = 333.3$ when $t = 0$ and $S = 938.2$ when $t = 8$. We use these two points to find the slope. Since $S = f(t)$, we know the slope is $\Delta S/\Delta t$:

$$m = \frac{\Delta S}{\Delta t} = \frac{938.2 - 333.3}{8 - 0} = 75.61.$$

The vertical intercept (when $t = 0$) is 333.3, so the linear equation is

$$S = 333.3 + 75.61t.$$

(b) The slope is 75.61 million CDs per year. Sales of music CDs were increasing at a rate of 75.61 million CDs a year. The vertical intercept is 333.3 million CDs, and represents the sales in the year 1991.

(c) In the year 2005, we have $t = 14$ and

$$\text{Sales } = 333.3 + 75.61 \cdot 14 = 1391.84 \text{ million CDs.}$$

29. **(a)** Given the two points $(0, 32)$ and $(100, 212)$, and assuming the graph in Figure 1.4 is a line,

$$\text{Slope} = \frac{212 - 32}{100} = \frac{180}{100} = 1.8.$$

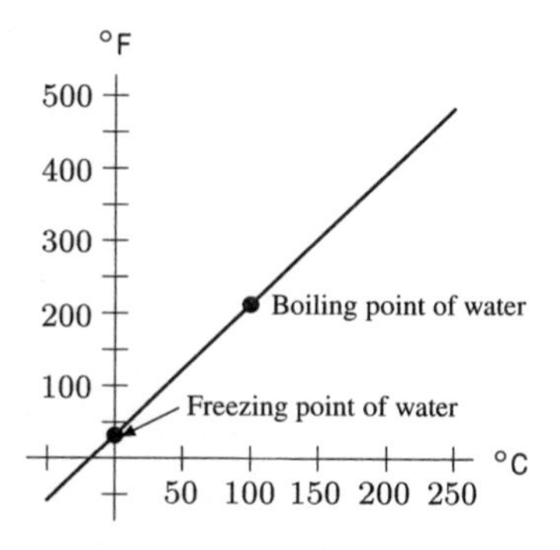

Figure 1.4

(b) The °F-intercept is $(0, 32)$, so

$$°\text{F} = 1.8(°\text{C}) + 32.$$

(c) If the temperature is 20°Celsius, then

$$°\text{F} = 1.8(20) + 32 = 68°\text{F}.$$

(d) If °F = °C, then

$$\begin{aligned} °\text{C} &= 1.8°\text{C} + 32 \\ -32 &= 0.8°\text{C} \\ °\text{C} &= -40° = °\text{F}. \end{aligned}$$

Solutions for Section 1.3

1. The graph shows a concave up function.

5. One possible answer is shown in Figure 1.5.

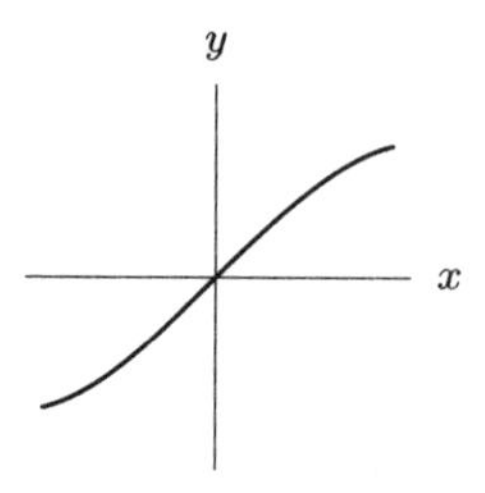

Figure 1.5

9. **(a)** The value of exports was higher in 1990 than in 1960. The 1960 figure looks like 600 and the 1990 figure looks like 3300, making the 1990 figure 2700 billion dollars higher than the 1960 figure.

(b) The average rate of change R is the change in values divided by the change in time.

$$\begin{aligned} R &= \frac{3300 - 600}{1990 - 1960} \\ &= \frac{2700}{30} \\ &= 90 \text{ billion dollars per year.} \end{aligned}$$

Between 1960 and 1990, the value of world exports has increased by an average of 90 billion dollars per year.

13. The average rate of change R between $x = 1$ and $x = 3$ is

$$\begin{aligned} R &= \frac{f(3) - f(1)}{3 - 1} \\ &= \frac{18 - 2}{2} \\ &= \frac{16}{2} \\ &= 8. \end{aligned}$$

17. When $t = 0$, we have $B = 1000(1.08)^0 = 1000$. When $t = 5$, we have $B = 1000(1.08)^5 = 1469.33$. We have

$$\text{Average rate of change } = \frac{\Delta B}{\Delta t} = \frac{1469.33 - 1000}{5 - 0} = 93.87 \text{ dollars/year.}$$

During this five year period, the amount of money in the account increased at an average rate of \$93.87 per year.

21. **(a)**

$$\begin{aligned}\text{Change between 1991 and 1995} &= \text{Sales in 1995} - \text{Sales in 1991}\\ &= 16{,}202 - 4779\\ &= 11{,}423 \text{ million dollars.}\end{aligned}$$

(b)

$$\begin{aligned}\begin{array}{c}\text{Average rate of change}\\ \text{between 1991 and 1995}\end{array} &= \frac{\text{Change in sales}}{\text{Change in time}}\\ &= \frac{\text{Sales in 1995} - \text{Sales in 1991}}{1995 - 1991}\\ &= \frac{16{,}202 - 4779}{1995 - 1991}\\ &= \frac{11{,}423}{4}\\ &= 2855.7 \text{ million dollars per year.}\end{aligned}$$

This means that Intel's sales increased on average by 2855.7 million dollars per year between 1991 and 1995.

(c) The average rate of change of sales between 1995 and 1997 is

$$\frac{25{,}070 - 16{,}202}{1997 - 1995} = \frac{8868}{2} = 4434 \text{ million dollars per year.}$$

If sales continue to increase at 4434 million dollars per year, then

$$\begin{aligned}\text{Sales in 1998} &= \text{Sales in 1997 } + 4434 = 29{,}504 \text{ million dollars per year.}\\ \text{Sales in 1999} &= \text{Sales in 1998 } + 4434 = 33{,}938 \text{ million dollars per year.}\\ \text{Sales in 2000} &= \text{Sales in 1999 } + 4434 = 38{,}372 \text{ million dollars per year.}\\ \text{Sales in 2001} &= \text{Sales in 2000 } + 4434 = 42{,}806 \text{ million dollars per year.}\end{aligned}$$

If Intel's sales continue to increase at the rate of 4434 million dollars per year, then sales will first reach 40,000 million dollars in the year 2001.

25.

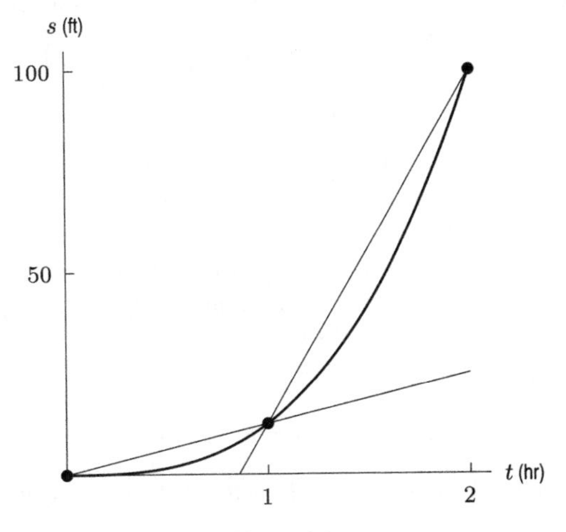

Figure 1.6

The graph in Figure 1.6 satisfies the properties. The graph is always increasing so has positive slope for all lines between any two of its points, making the average velocity positive for all intervals. Furthermore, the line corresponding to the interval from $t = 1$ to $t = 2$ has a greater slope than the interval from $t = 0$ to $t = 1$, making the average velocity of the first half less than that of the second half of the trip.

29. **(a)** Advertising is generally cheaper in bulk; spending more money will give better and better marginal results initially, hence concave up. (Spending \$5,000 could give you a big newspaper ad reaching 200,000 people; spending \$100,000 could give you a series of TV spots reaching 50,000,000 people.) A graph is shown below, left. But after a certain point, you may "saturate" the market, and the increase in revenue slows down – hence concave down.

(b) The temperature of a hot object decreases at a rate proportional to the difference between its temperature and the temperature of the air around it. Thus, the temperature of a very hot object decreases more quickly than a cooler object. The graph is decreasing and concave up. (We are assuming that the coffee is all at the same temperature.)

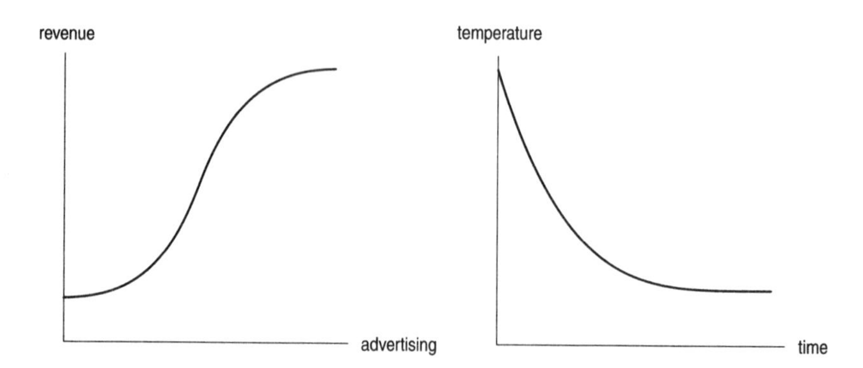

Solutions for Section 1.4

1. **(a)** We know that regardless of the number of rides one takes, one must pay \$7 to get in. After that, for each ride you must pay another \$1.50, thus the function $R(n)$ is

$$R(n) = 7 + 1.5n.$$

(b) Substituting in the values $n = 2$ and $n = 8$ into our formula for $R(n)$ we get

$$R(2) = 7 + 1.5(2) = 7 + 3 = \$10.$$

This means that admission and 2 rides costs \$10.

$$R(8) = 7 + 1.5(8) = 7 + 12 = \$19.$$

This means that admission and 8 rides costs \$19.

5. **(a)** The company makes a profit when the revenue is greater than the cost of production. Looking at the figure in the problem, we see that this occurs whenever more than roughly 335 items are produced and sold.

(b) If $q = 600$, revenue ≈ 2400 and cost ≈ 1750 so profit is about \$650.

9. **(a)** We have $C(q) = 6000 + 2q$ and $R(q) = 5q$ and so

$$\pi(q) = R(q) - C(q) = 5q - (6000 + 2q) = -6000 + 3q.$$

(b) See Figure 1.7. We find the break-even point, q_0, by setting the revenue equal to the cost and solving for q:

$$\begin{aligned} \text{Revenue} &= \text{Cost} \\ 5q &= 6000 + 2q \\ 3q &= 6000 \\ q &= 2000. \end{aligned}$$

The break-even point is $q_0 = 2000$ puzzles. Notice that this is the same answer we get if we set the profit function equal to zero.

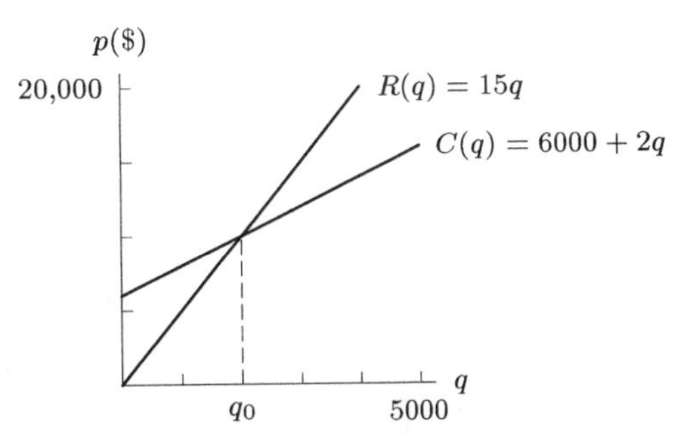

Figure 1.7: Cost and revenue functions for the jigsaw puzzle company

13. (a) We know that the fixed cost of the first price list is \$100 and the variable cost is \$0.03. Thus, the cost of making q copies under the first option is

$$C_1(q) = 100 + 0.03q.$$

We know that the fixed cost of the second price list is \$200 and the variable cost is \$0.02. Thus, the cost of making q copies under the second option is

$$C_2(q) = 200 + 0.02q.$$

(b) At 5000 copies, the first price list gives the cost

$$C_1(5000) = 100 + 5000(0.03) = 100 + 150 = \$250.$$

At 5000 copies, the second price list gives the cost

$$C_2(5000) = 200 + 5000(0.02) = 200 + 100 = \$300.$$

Thus, for 5000 copies, the first price list is cheaper.

(c) We are asked to find the point q at which

$$C_1(q) = C_2(q).$$

Solving we get

$$\begin{aligned} C_1(q) &= C_2(q) \\ 100 + 0.03q &= 200 + 0.02q \\ 100 &= 0.01q \\ q &= 10{,}000. \end{aligned}$$

Thus, if one needs to make ten thousand copies, the cost under both price lists will be the same.

17. (a) We know that the total amount the company will spend on raw materials will be

$$\text{price for raw materials} = \$100 \cdot m$$

where m is the number of units of raw materials the company will buy. We know that the total amount the company will spend on paying employees will be

$$\text{total employee expenditure} = \$25{,}000 \cdot r$$

where r is the number of employees the company will hire. Since the total amount the company spends is \$500,000, we get

$$25{,}000r + 100m = 500{,}000.$$

(b) Solving for m we get

$$\begin{aligned} 25{,}000r + 100m &= 500{,}000 \\ 100m &= 500{,}000 - 25{,}000r \\ m &= 5000 - 250r. \end{aligned}$$

(c) Solving for r we get

$$\begin{aligned} 25{,}000r + 100m &= 500{,}000 \\ 25{,}000r &= 500{,}000 - 100m \\ r &= 20 - \frac{100}{25{,}000}m \\ &= 20 - \frac{1}{250}m. \end{aligned}$$

21. **(a)** We know that as the price per unit increases, the quantity supplied increases, while the quantity demanded decreases. So Table 1.22 of the text is the demand curve (since as the price increases the quantity decreases), while Table 1.23 of the text is the supply curve (since as the price increases the quantity increases.)

(b) Looking at the demand curve data in Table 1.22 we see that a price of \$155 gives a quantity of roughly 14.

(c) Looking at the supply curve data in Table 1.23 we see that a price of \$155 gives a quantity of roughly 24.

(d) Since supply exceeds demand at a price of \$155, the shift would be to a lower price.

(e) Looking at the demand curve data in Table 1.22 we see that if the price is less than or equal to \$143 the consumers would buy at least 20 items.

(f) Looking at the data for the supply curve (Table 1.23) we see that if the price is greater than or equal to \$110 the supplier will produce at least 20 items.

25. **(a)** $k = p_1 s + p_2 l$ where s = # of liters of soda and l = # of liters of oil.

(b) If $s = 0$, then $l = k/p_2$. Similarly, if $l = 0$, then $s = k/p_1$. These two points give you enough information to draw a line containing the points which satisfy the equation.

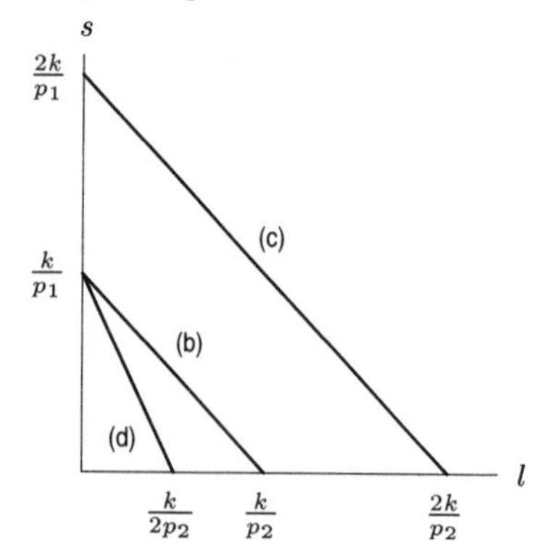

(c) If the budget is doubled, we have the constraint: $2k = p_1 s + p_2 l$. We find the intercepts as before. If $s = 0$, then $l = 2k/p_2$; if $l = 0$, then $s = 2k/p_1$. The intercepts are both twice what they were before.

(d) If the price of oil doubles, our constraint is $k = p_1 s + 2p_2 l$. Then, calculating the intercepts gives that the s intercept remains the same, but the l intercept gets cut in half. In other words, $s = 0$ means $l = k/(2p_2) = (1/2)k/p_2$. Therefore the maximum amount of oil you can buy is half of what it was previously.

29. We know that at the point where the price is \$1 per scoop the quantity must be 240. Thus we can fill in the graph as follows:

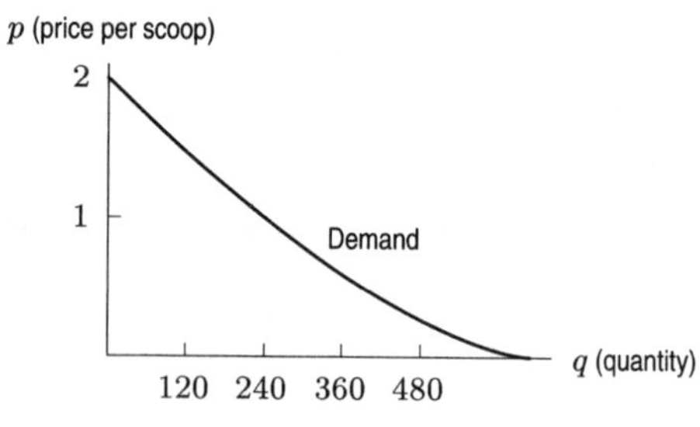

Figure 1.8

(a) Looking at Figure 1.8 we see that when the price per scoop is half a dollar, the quantity given by the demand curve is roughly 360 scoops.

(b) Looking at Figure 1.8 we see that when the price per scoop is \$1.50, the quantity given by the demand curve is roughly 120 scoops.

Solutions for Section 1.5

1. **(a)** Initial amount $= 100$; exponential growth; growth rate $= 7\% = 0.07$.
   (b) Initial amount $= 5.3$; exponential growth; growth rate $= 5.4\% = 0.054$.
   (c) Initial amount $= 3500$; exponential decay; decay rate $= -7\% = -0.07$.
   (d) Initial amount $= 12$; exponential decay; decay rate $= -12\% = -0.12$.

5. **(a)** Since the percent rate of change is constant, A is an exponential function of t. The initial quantity is 50 and the base is $1 - 0.06 = 0.94$. The formula is

$$A = 50(0.94)^t.$$

   (b) When $t = 24$, we have $A = 50(0.94)^{24} = 11.33$. After one day, 11.33 mg of quinine remains.
   (c) See Figure 1.9. The graph is decreasing and concave up, as we would expect with an exponential decay function.
   (d) In Figure 1.9, it appears that when $A = 5$, we have $t \approx 37$. It takes about 37 hours until 5 mg remains.

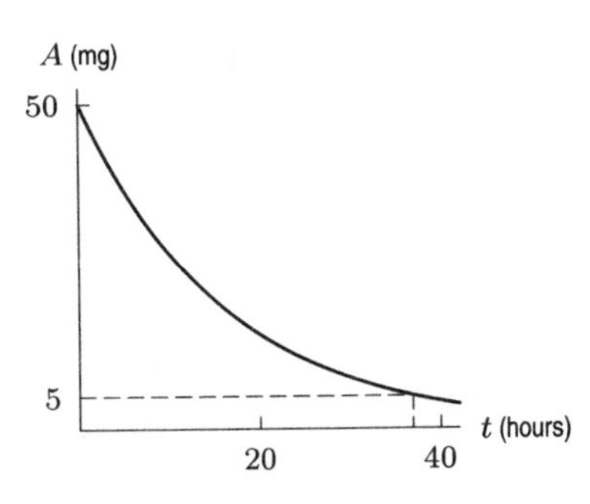

Figure 1.9

9. Since we are looking for an annual percent increase, we use an exponential function. If V represents the value of the company, and t represent the number of years since it was started, we have $V = Ca^t$. The initial value of the company was \$4000, so $V = 4000a^t$. To find a, we use the fact that $V = 14{,}000{,}000$ when $t = 40$:

$$\begin{aligned} V &= 4000a^t \\ 14{,}000{,}000 &= 4000a^{40} \\ 3500 &= a^{40} \\ a &= (3500)^{1/40} = 1.226. \end{aligned}$$

   The value of the company increased at an average rate of 22.6% per year.

13. This looks like an exponential decay function $y = Ca^t$. The y-intercept is 30 so we have $y = 30a^t$. We use the point $(25, 6)$ to find a:

$$\begin{aligned} y &= 30a^t \\ 6 &= 30a^{25} \\ 0.2 &= a^{25} \\ a &= (0.2)^{1/25} = 0.94. \end{aligned}$$

   The formula is $y = 30(0.94)^t$.

17. Since the decay rate is 2.9%, the base of the exponential function is $1 - 0.029 = 0.971$. If P represents the percent of forested land in 1980 remaining t years after 1980, when $t = 0$ we have $P = 100$. The formula is $P = 100(0.971)^t$. The percent remaining in 1990 is $P = 100(0.971)^{10} = 74.5$. About 74.5% of the land was still covered in forests 10 years later.

21. See Figure 1.10. The graph is decreasing and concave up. It looks like an exponential decay function with y-intercept 100.

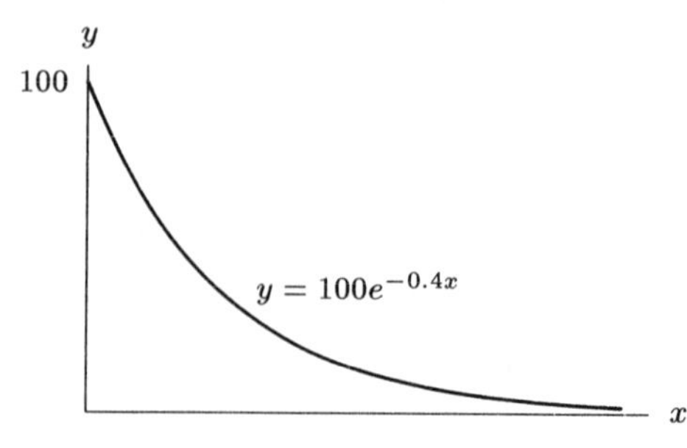

Figure 1.10

25. (a) Since the growth is exponential, the equation is of the form $P(t) = P_0 a^t$, where $P_0 = 5.6$, the population at time $t = 0$. Since the growth rate is 1.2%, $a = 1.012$. Thus, $P(t) = 5.6(1.012)^t$, in billions.

(b)

$$\begin{aligned}\text{Average rate of change} &= \frac{P(6) - P(0)}{6 - 0}\\ &= \frac{6.015 - 5.6}{6}\\ &= 0.069 \text{ billion people per year increase.}\end{aligned}$$

(c)

$$\begin{aligned}\text{Average rate of change} &= \frac{P(26) - P(16)}{26 - 16}\\ &= \frac{7.636 - 6.778}{10}\\ &= 0.086 \text{ billion people per year increase.}\end{aligned}$$

Solutions for Section 1.6

1. Taking natural logs of both sides we get

$$\ln(5^t) = \ln 7.$$

This gives

$$t \ln 5 = \ln 7$$

or in other words

$$t = \frac{\ln 7}{\ln 5} \approx 1.209.$$

5. Dividing both sides by 25 we get

$$4 = 1.5^t.$$

Taking natural logs of both side gives

$$\ln(1.5^t) = \ln 4.$$

This gives

$$t \ln 1.5 = \ln 4$$

or in other words

$$t = \frac{\ln 4}{\ln 1.5} \approx 3.419.$$

9. Dividing both sides by 2 we get

$$2.5 = e^t.$$

Taking the natural log of both sides gives

$$\ln(e^t) = \ln 2.5.$$

This gives

$$t = \ln 2.5 \approx 0.9163.$$

13. Dividing both sides by P we get

$$\frac{B}{P} = e^{rt}.$$

Taking the natural log of both sides gives

$$\ln(e^{rt}) = \ln\left(\frac{B}{P}\right).$$

This gives

$$rt = \ln\left(\frac{B}{P}\right) = \ln B - \ln P.$$

Dividing by r gives

$$t = \frac{\ln B - \ln P}{r}.$$

17. Initial quantity = 5; growth rate = 0.07 = 7%.

21. Since $e^{0.08} = 1.0833$, and $e^{-0.3} = 0.741$, we have

$$P = e^{0.08t} = \left(e^{0.08}\right)^t = (1.0833)^t \text{ and}$$
$$Q = e^{-0.3t} = \left(e^{-0.3}\right)^t = (0.741)^t.$$

25. **(a)** The continuous percent growth rate is 15%.

(b) We want $P_0 a^t = 10e^{0.15t}$, so we have $P_0 = 10$ and $a = e^{0.15} = 1.162$. The corresponding function is

$$P = 10(1.162)^t.$$

(c) Since the base in the answer to part (b) is 1.162, the annual percent growth rate is 16.2%. This annual rate is equivalent to the continuous growth rate of 15%.

(d) When we sketch the graphs of $P = 10e^{0.15t}$ and $P = 10(1.162)^t$ on the same axes, we only see one graph. These two exponential formulas are two ways of representing the same function, so the graphs are the same. See Figure 1.11.

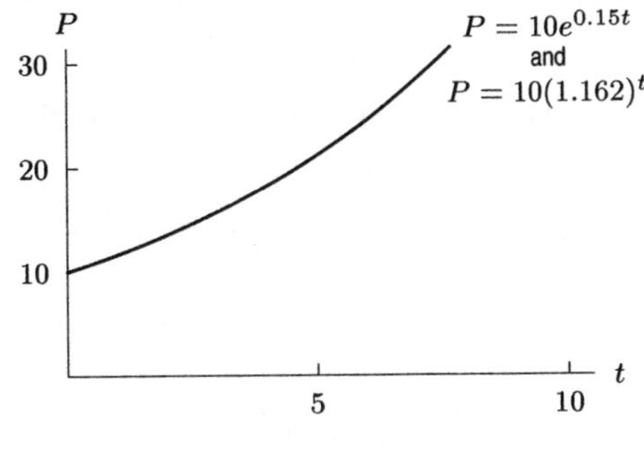

Figure 1.11

29. $P = 7(e^{-\pi})^t = 7(0.0432)^t$. Exponential decay because $-\pi < 0$ or $0.0432 < 1$.

33. Since we want $(0.55)^t = e^{kt} = (e^k)^t$, so $0.55 = e^k$, and $k = \ln 0.55 = -0.5978$. Thus $P = 4e^{-0.5978t}$. Since -0.5978 is negative, this represents exponential decay.

37. **(a)** $P = 3.6(1.034)^t$

(b) Since $P = 3.6e^{kt} = 3.6(1.034)^t$, we have

$$\begin{aligned} e^{kt} &= (1.034)^t \\ kt &= t\ln(1.034) \\ k &= 0.0334. \end{aligned}$$

Thus, $P = 3.6e^{0.0334t}$.

(c) The annual growth rate is 3.4%, while the continuous growth rate is 3.3%. Thus the growth rates are not equal, though for small growth rates (such as these), they are close. The annual growth rate is larger.

41. We find k with $e^{kt} = (1.10)^t$. We have $e^k = 1.10$ so $k = \ln(1.10) = 0.0953$. The corresponding continuous percent growth rate is 9.53%.

Solutions for Section 1.7

1. We know that the formula for the total process is

$$P(t) = P_0 a^t$$

where P_0 is the initial size of the process and $a = 1 + 0.07 = 1.07$. We are asked to find the doubling time. Thus, we solve

$$\begin{aligned} 2P_0 &= P_0(1.07)^t \\ 1.07^t &= 2 \\ \ln(1.07^t) &= \ln 2 \\ t\ln 1.07 &= \ln 2 \\ t &= \frac{\ln 2}{\ln 1.07} \approx 10.24. \end{aligned}$$

The doubling time is roughly 10.24 years.

Alternatively we could have used the "rule of 70" to estimate that the doubling time is approximately

$$\frac{70}{7} = 10 \text{ years.}$$

5. We use the equation $B = Pe^{rt}$, where P is the initial principal, t is the time in years the deposit is in the account, r is the annual interest rate and B is the balance. After $t = 5$ years, we will have

$$B = (10{,}000)e^{(0.08)5} = (10{,}000)e^{0.4} \approx \$14{,}918.25.$$

9. We know that the formula for the balance after t years in an account which is compounded continuously is

$$P(t) = P_0 e^{rt}$$

where P_0 is the initial deposit and r is the interest rate. In our case we are told that the rate is $r = 0.08$, so we have

$$P(t) = P_0 e^{0.08t}.$$

We are asked to find the value for P_0 such that after three years the balance would be \$10,000. That is, we are asked for P_0 such that

$$10{,}000 = P_0 e^{0.08(3)}.$$

Solving we get

$$\begin{aligned} 10{,}000 &= P_0 e^{0.08(3)} \\ &= P_0 e^{0.24} \\ &\approx P_0(1.27125) \\ P_0 &\approx \frac{10{,}000}{1.27125} \\ &\approx 7866.28. \end{aligned}$$

Thus, \$7866.28 should be deposited into this account so that after three years the balance would be \$10,000.

13. **(a)** Since $P(t)$ is an exponential function, it will be of the form $P(t) = P_0 a^t$, where P_0 is the initial population and a is the base. $P_0 = 200$, and a 5% growth rate means $a = 1.05$. Thus, we get

$$P(t) = 200(1.05)^t.$$

(b) The graph is shown in Figure 1.12.

(c) Evaluating gives that $P(10) = 200(1.05)^{10} \approx 326$.

(d) From the graph we see that the population is 400 at about $t = 15$, so the doubling time appears to be about 15 years.

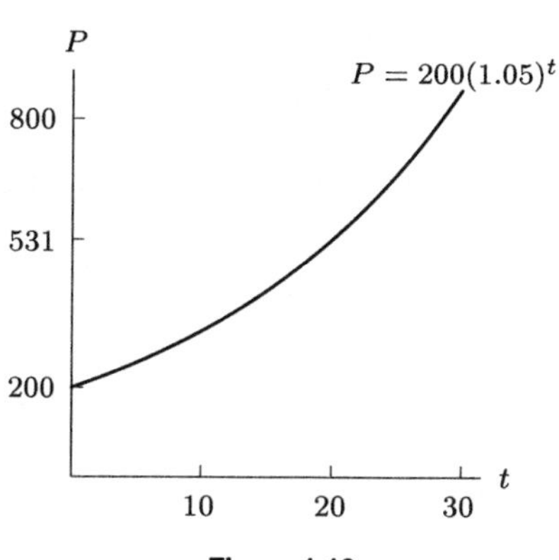

Figure 1.12

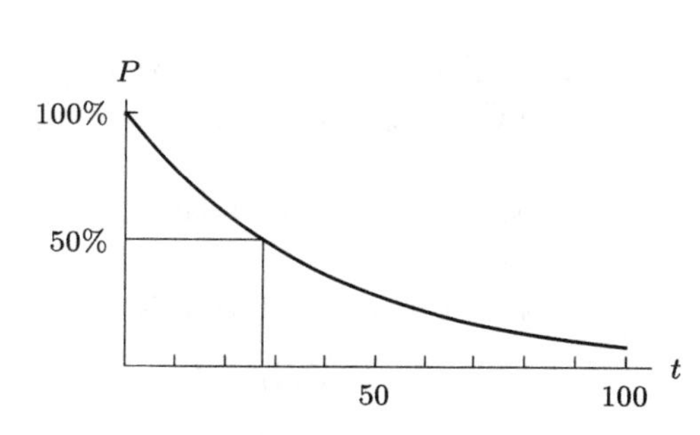

Figure 1.13

17. **(a)** Since $P(t)$ is an exponential function, it will be of the form $P(t) = P_0 a^t$. We have $P_0 = 1$, since 100% is present at time $t = 0$, and $a = 0.975$, because each year 97.5% of the contaminant remains. Thus,

$$P(t) = (0.975)^t.$$

(b) The graph is shown in Figure 1.13.

(c) The half-life is about 27 years, since $(0.975)^{27} \approx 0.5$.

(d) At time $t = 100$ there is about 8% remaining, since $(0.975)^{100} \approx 0.08$.

21. Given the doubling time of 5 hours, we can solve for the bacteria's growth rate;

$$2P_0 = P_0 e^{k5}$$
$$k = \frac{\ln 2}{5}.$$

So the growth of the bacteria population is given by:

$$P = P_0 e^{\ln(2)t/5}.$$

We want to find t such that

$$3P_0 = P_0 e^{\ln(2)t/5}.$$

Therefore we cancel P_0 and apply ln. We get

$$t = \frac{5\ln(3)}{\ln(2)} = 7.925 \text{ hours}.$$

25. **(a)** (i) We know that the formula for the account balance at time t is given by

$$P(t) = P_0 e^{rt}$$

(since the account is being compounded continuously), where P_0 is the initial deposit and r is the annual rate. In our case the initial deposit is

$$P_0 = 24$$

and the annual rate is

$$r = 0.05.$$

Thus, the formula for the balance is

$$P(t) = 24e^{0.05t}.$$

We are asked for the amount of money in the account in the year 2000, that is we are asked for the amount of money in the account 374 years after the initial deposit. This gives

$$\begin{aligned}\text{amount of money in the account in the year 2000} &= P(374)\\ &= 24e^{0.05(374)}\\ &= 24e^{18.7}\\ &\approx 24(132{,}223{,}000)\\ &= 3{,}173{,}352{,}000.\end{aligned}$$

Thus, in the year 2000, there would be about 3.17 billion dollars in the account.

(ii) Here the annual rate is

$$r = 0.07.$$

Thus, the formula for the balance is

$$P(t) = 24e^{0.07t}.$$

Again, we want $P(374)$. This gives

$$\begin{aligned}\text{Amount of money in the account in the year 2000} &= P(374)\\ &= 24e^{0.07(374)}\\ &= 24e^{26.18}\\ &\approx 24(234{,}330{,}890{,}000)\\ &= 5{,}623{,}941{,}360{,}000.\end{aligned}$$

Thus, in the year 2000, there would be about 5.62 trillion dollars in the account.

(b) Here the annual rate is

$$r = 0.06.$$

Thus, the formula for the balance is

$$P(t) = 24e^{0.06t}.$$

We are asked to solve for t such that

$$P(t) = 1{,}000{,}000.$$

Solving we get

$$\begin{aligned}1{,}000{,}000 &= B(t) = 24e^{0.06t}\\ \frac{1{,}000{,}000}{24} &= e^{0.06t}\\ e^{0.06t} &\approx 41{,}666.7\\ \ln e^{0.06t} &\approx \ln 41{,}666.7\\ 0.06t &\approx \ln 41{,}666.7\\ t &\approx \frac{\ln 41{,}666.7}{0.06} \approx 177.3.\end{aligned}$$

Thus, there would be one million dollars in the account 177.3 years after the initial deposit, that is roughly in 1803.

29. We use $P = P_0e^{kt}$. Since 200 grams are present initially, we have $P = 200e^{kt}$. To find k, we use the fact that the half-life is 8 years:

$$\begin{aligned}P &= 200e^{kt}\\ 100 &= 200e^{k(8)}\\ 0.5 &= e^{8k}\\ \ln(0.5) &= 8k\\ k &= \frac{\ln(0.5)}{8} \approx -0.087.\end{aligned}$$

We have

$$P = 200e^{-0.087t}.$$

The amount remaining after 12 years is

$$P = 200e^{-0.087(12)} = 70.4 \text{ grams}.$$

To find the time until 10% remains, we use $P = 0.10P_0 = 0.10(200) = 20$:

$$\begin{aligned} 20 &= 200e^{-0.087t} \\ 0.10 &= e^{-0.087t} \\ \ln(0.10) &= -0.087t \\ t &= \frac{\ln(0.10)}{-0.087} = 26.5 \text{ years}. \end{aligned}$$

33. **(a)** The total present value for each of the two choices are in the following table. Choice 1 is the preferred choice since it has the larger present value.

Choice 1			Choice 2	
Year	Payment	Present value	Payment	Present value
0	1500	1500	1900	1900
1	3000	$3000/(1.05) = 2857.14$	2500	$2500/(1.05) = 2380.95$
	Total	4357.14	Total	4280.95

(b) The difference between the choices is an extra \$400 now (\$1900 in Choice 2 instead of \$1500 in Choice 1) versus an extra \$500 one year from now (\$3000 in Choice 1 instead of \$2500 in Choice 2). Since $400 \times 1.25 = 500$, Choice 2 is better at interest rates above 25%.

37. **(a)** The following table gives the present value of the cash flows. The total present value of the cash flows is \$116,224.95.

Year	Payment	Present value
1	50,000	$50{,}000/(1.075) = 46,511.63$
2	40,000	$40{,}000/(1.075)^2 = 34,613.30$
3	25,000	$25{,}000/(1.075)^3 = 20,124.01$
4	20,000	$20{,}000/(1.075)^4 = 14,976.01$
	Total	$116,224.95$

(b) The present value of the cash flows, \$116,224.95, is larger than the price of the machine, \$97,000, so the recommendation is to buy the machine.

Solutions for Section 1.8

1. **(a)** $g(2+h) = (2+h)^2 + 2(2+h) + 3 = 4 + 4h + h^2 + 4 + 2h + 3 = h^2 + 6h + 11$.
(b) $g(2) = 2^2 + 2(2) + 3 = 4 + 4 + 3 = 11$, which agrees with what we get by substituting $h = 0$ into (a).
(c) $g(2+h) - g(2) = (h^2 + 6h + 11) - (11) = h^2 + 6h$.

5. **(a)** $g(h(x)) = g(x^3 + 1) = \sqrt{x^3 + 1}$.
(b) $h(g(x)) = h(\sqrt{x}) = (\sqrt{x})^3 + 1 = x^{3/2} + 1$.
(c) $h(h(x)) = h(x^3 + 1) = (x^3 + 1)^3 + 1$.
(d) $g(x) + 1 = \sqrt{x} + 1$.
(e) $g(x + 1) = \sqrt{x + 1}$.

9.

(a)

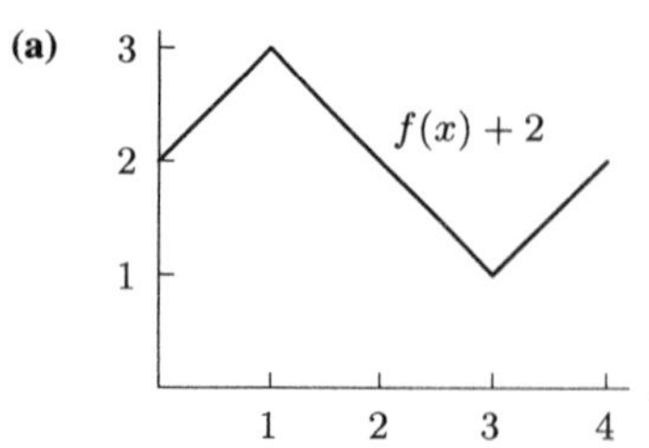

(b)

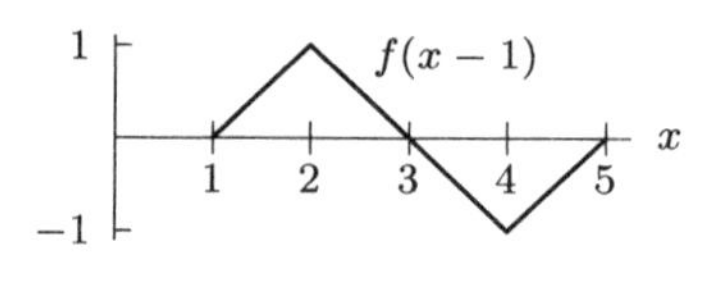

(c)

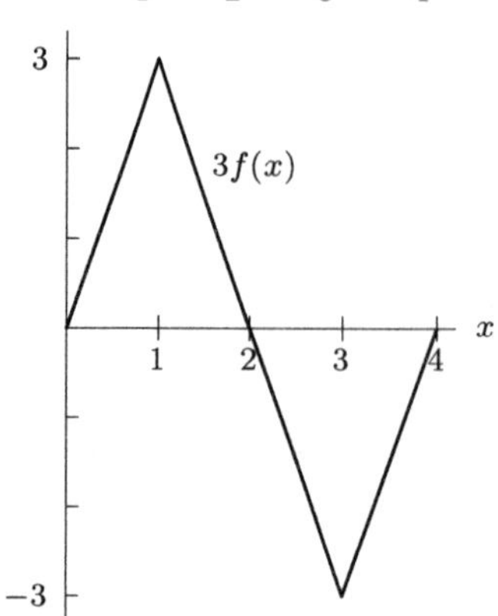

(d)

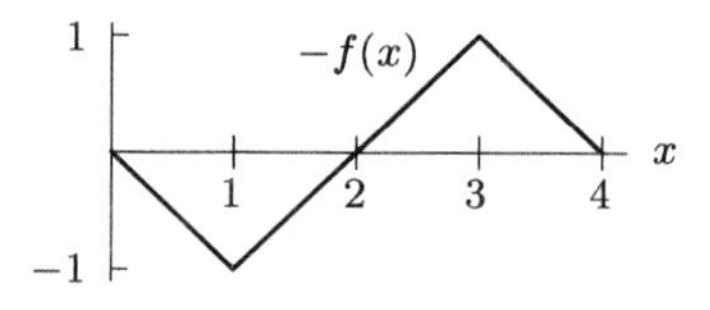

13.

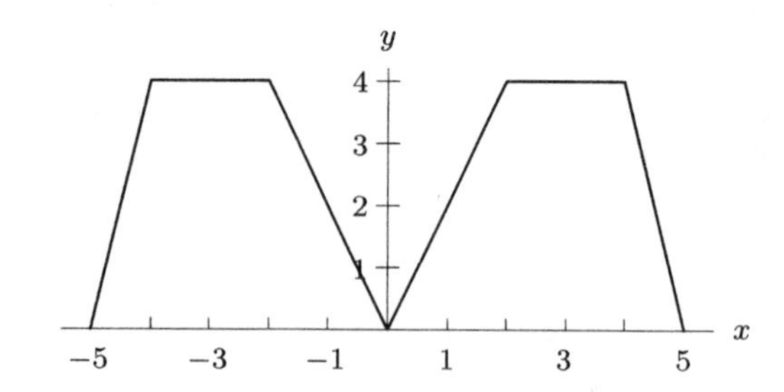

Figure 1.14: Graph of $y = 2f(x)$

17.

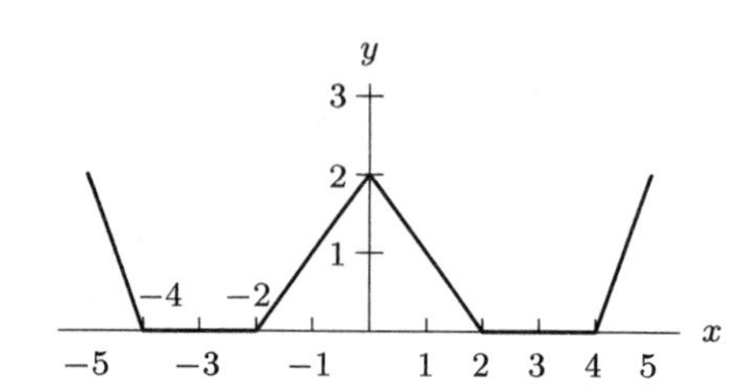

Figure 1.15: Graph of $2 - f(x)$

21. $f(g(1)) = f(2) \approx 0.4$.

25. **(a)** $f(n) + g(n) = (3n^2 - 2) + (n + 1) = 3n^2 + n - 1$.

(b) $f(n)g(n) = (3n^2 - 2)(n + 1) = 3n^3 + 3n^2 - 2n - 2$.

(c) The domain of $f(n)/g(n)$ is defined everywhere where $g(n) \neq 0$, i.e. for all $n \neq -1$.

(d) $f(g(n)) = 3(n + 1)^2 - 2 = 3n^2 + 6n + 1$.

(e) $g(f(n)) = (3n^2 - 2) + 1 = 3n^2 - 1$.

29. $m(z + h) - m(z - h) = (z + h)^2 - (z - h)^2 = z^2 + 2hz + h^2 - (z^2 - 2hz + h^2) = 4hz$.

33. To find $f(g(-3))$, we notice that $g(-3) = 3$ and $f(3) = 0$ so $f(g(-3)) = f(3) = 0$. We find the other values similarly. See Table 1.1. The graphs are shown in Figure 1.16.

Table 1.1

x	-3	-2	-1	0	1	2	3
$f(g(x))$	0	1	2	3	2	1	0
$g(f(x))$	0	-1	-2	-3	-2	-1	0

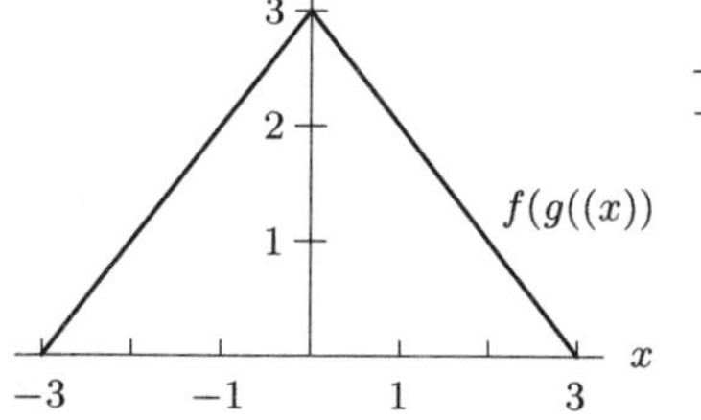

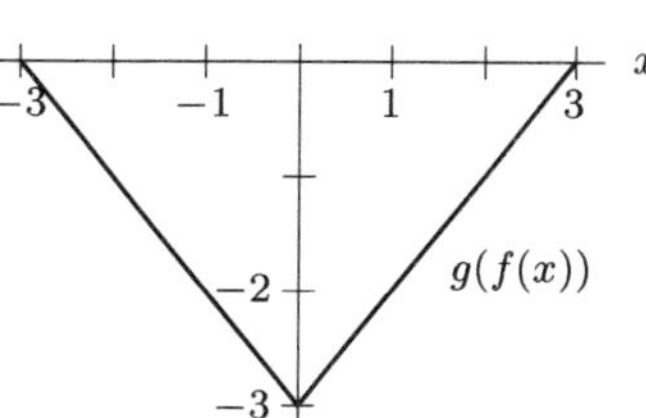

Figure 1.16

Solutions for Section 1.9

1. $y = 3x^{-2}$; $k = 3, p = -2$.
5. $y = \frac{5}{2}x^{-1/2}$; $k = \frac{5}{2}, p = -1/2$.
9. $y = 5^3 \cdot x^3 = 125x^3$; $k = 125, p = 3$.
13. We know that E is proportional to v^3, so $E = kv^3$, for some constant k.
17. If p is proportional to t, then $p = kt$ for some fixed constant k. From the values $t = 10, p = 25$, we have $25 = k(10)$, so $k = 2.5$. To see if p is proportional to t, we must see if $p = 2.5t$ gives all the values in the table. However, when we check the values $t = 20, p = 60$, we see that $60 \neq 2.5(20)$. Thus, p is not proportional to t.
21. **(a)** The degree of $x^2 + 10x - 5$ is 2 and its leading coefficient is positive.
    (b) In a large window, the function looks like x^2. See Figure 1.17.

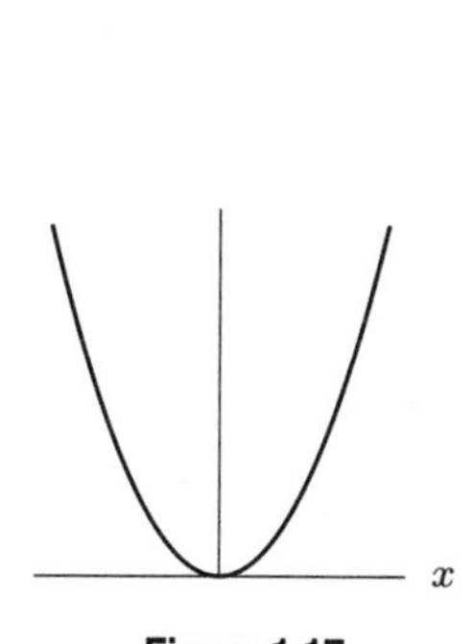

Figure 1.17

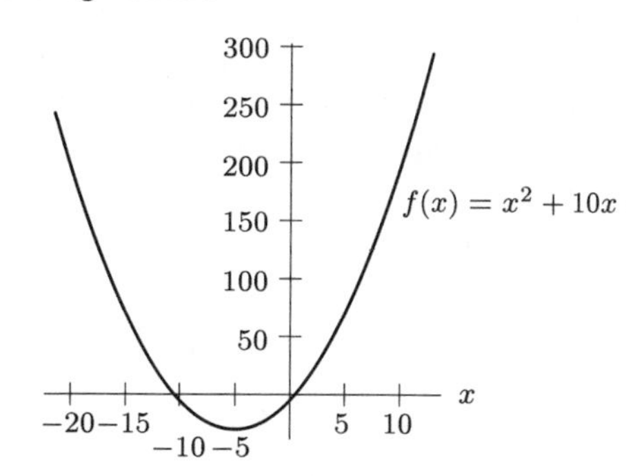

Figure 1.18

(c) The function has one turning point which is what we expect of a quadratic. See Figure 1.18.

25. **(a)** The degree of $-9x^5 + 72x^3 + 12x^2$ is 5 and its leading coefficient is negative.
    (b) In a large window, the graph of $f(x)$ looks like $-9x^5$. See Figure 1.19.

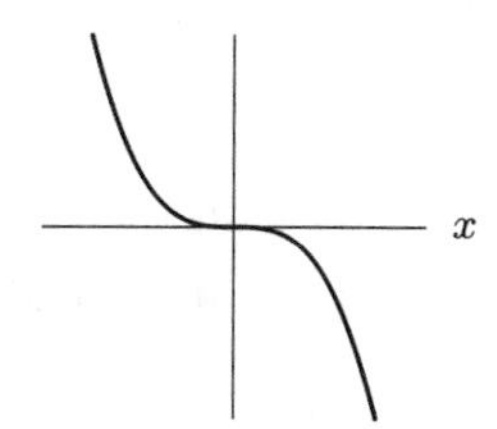

Figure 1.19

(c) It is very hard to see what's going on with this function by looking at just one graph. If we look at the graph of $f(x)$ from $x = -4$ to $x = 4$ in Figure 1.20, we see two obvious turning points but can't tell what's going on near the origin. If we look at the graph from $x = -0.5$ to $x = 0.5$ in Figure 1.21, we can see what's going on near the origin but we can no longer see the two turning points we saw in the larger scale plot. But we see that there are two turning points near the origin as well. Thus, there are four turning points total. This is one less than the degree of the polynomial, which is the maximum it could be.

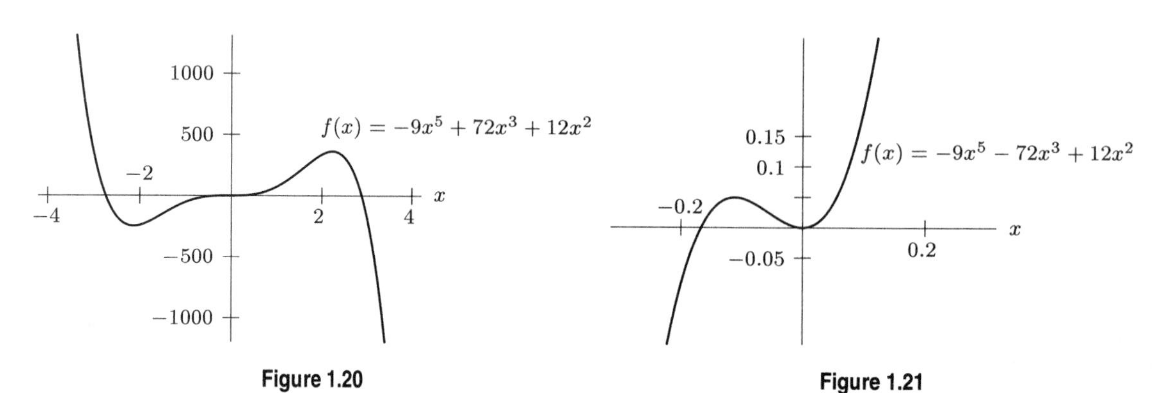

Figure 1.20

Figure 1.21

29. If $y = k \cdot x^3$, then we have $y/x^3 = k$ with k a constant. In other words, the ratio

$$\frac{y}{x^3} = \frac{\text{Weight}}{(\text{Length})^3}$$

should all be approximately equal (and the ratio will be the constant of proportionality k). For the first fish, we have

$$\frac{y}{x^3} = \frac{332}{(33.5)^3} = 0.0088.$$

If we check all 11 data points, we get the following values of the ratio y/x^3: 0.0088, 0.0088, 0.0087, 0.0086, 0.0086, 0.0088, 0.0087, 0.0086, 0.0087, 0.0088, 0.0088.These numbers are indeed approximately constant with an average of about 0.0087. Thus, $k \approx 0.0087$ and the allometric equation $y = 0.0087x^3$ fits this data well.

33. **(a)** $R(P) = kP(L - P)$, where k is a positive constant.

(b) A possible graph is in Figure 1.22.

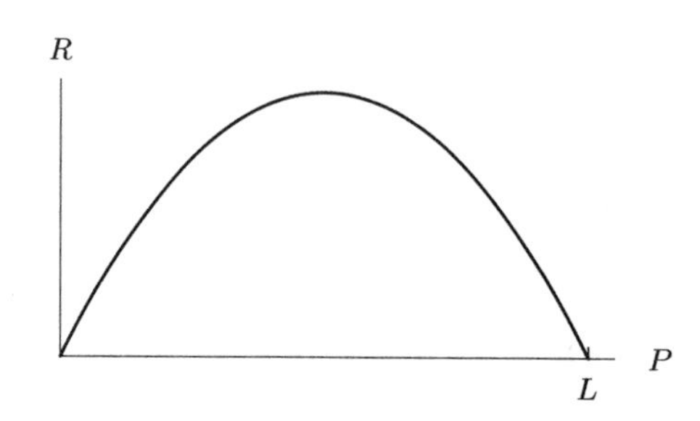

Figure 1.22

37. **(a)** We know that

$$q = 3000 - 20p,$$
$$C = 10{,}000 + 35q,$$

and

$$R = pq.$$

Thus, we get

$$\begin{aligned} C &= 10{,}000 + 35q \\ &= 10{,}000 + 35(3000 - 20p) \\ &= 10{,}000 + 105{,}000 - 700p \\ &= 115{,}000 - 700p \end{aligned}$$

and

$$\begin{aligned} R &= pq \\ &= p(3000 - 20p) \\ &= 3000p - 20p^2. \end{aligned}$$

(b) The graph of the cost and revenue functions are shown in Figure 1.23.

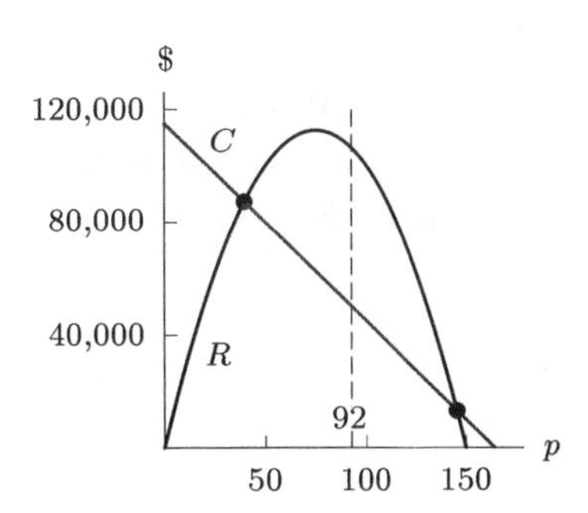

Figure 1.23

(c) It makes sense that the revenue function looks like a parabola, since if the price is too low, although a lot of people will buy the product, each unit will sell for a pittance and thus the total revenue will be low. Similarly, if the price is too high, although each sale will generate a lot of money, few units will be sold because consumers will be repelled by the high prices. In this case as well, total revenue will be low. Thus, it makes sense that the graph should rise to a maximum profit level and then drop.

(d) The club makes a profit whenever the revenue curve is above the cost curve in Figure 1.23. Thus, the club makes a profit when it charges roughly between \$40 and \$145.

(e) We know that the maximal profit occurs when the difference between the revenue and the cost is the greatest. Looking at Figure 1.23, we see that this occurs when the club charges roughly \$92.

Solutions for Section 1.10

1. It makes sense that sunscreen sales would be lowest in the winter (say from December through February) and highest in the summer (say from June to August). It also makes sense that sunscreen sales would depend almost completely on time of year, so the function should be periodic with a period of 1 year.

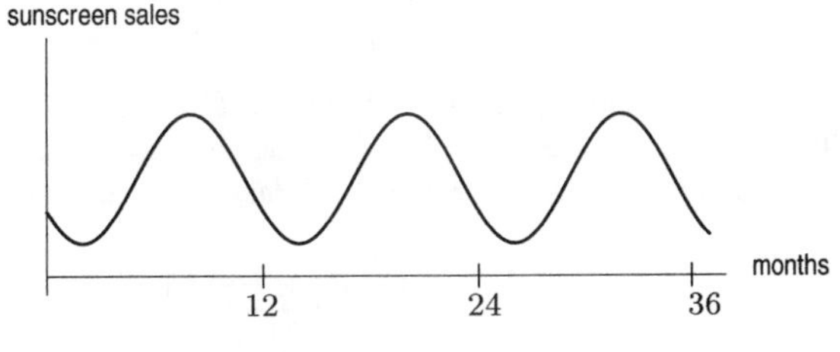

Figure 1.24

5. If we graph the data, it certainly looks like there's a pattern from $t = 20$ to $t = 45$ which begins to repeat itself after $t = 45$. And the fact that $f(20) = f(45)$ helps to reinforce this theory.

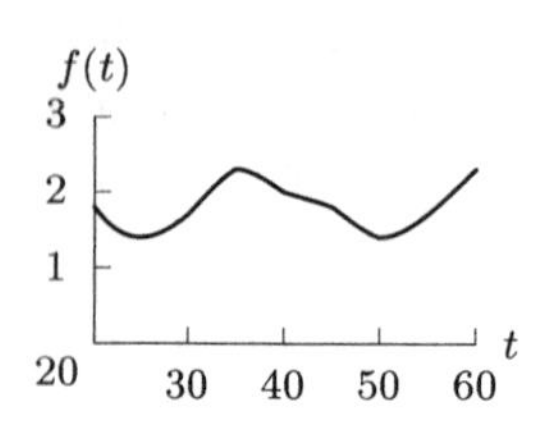

Figure 1.25

Since it looks like the graph repeats after 45, the period is $(45 - 20) = 25$. The amplitude $= \frac{1}{2}(\max - \min) = \frac{1}{2}(2.3 - 1.4) = 0.45$. Assuming periodicity,

$$\begin{aligned} f(15) &= f(15+25) = f(40) = 2.0 \\ f(75) &= f(75-25) = f(50) = 1.4 \\ f(135) &= f(135-25-25-25) = f(60) = 2.3. \end{aligned}$$

9.

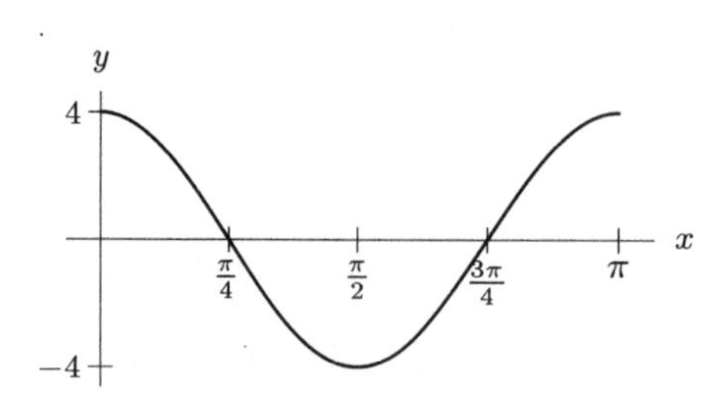

Figure 1.26

See Figure 1.26. The amplitude is 4; the period is π.

13. The levels of both hormones certainly look periodic. In each case, the period seems to be about 28 days. Estrogen appears to peak around the 12th day. Progesterone appears to peak from the 17th through 21st day. (Note: the days given in the answer are approximate so your answer may differ slightly.)

17. This graph is a cosine curve with period 6π and amplitude 5, so it is given by $f(x) = 5\cos\left(\frac{x}{3}\right)$.

21. This graph is an inverted cosine curve with amplitude 8 and period 20π, so it is given by $f(x) = -8\cos\left(\frac{x}{10}\right)$.

25. This graph has period 8, amplitude 3, and a vertical shift of 3 with no horizontal shift. It is given by

$$f(x) = 3 + 3\sin\left(\frac{2\pi}{8}x\right) = 3 + 3\sin\left(\frac{\pi}{4}x\right).$$

29. We use a cosine of the form

$$H = A\cos(Bt) + C$$

and choose B so that the period is 24 hours, so $2\pi/B = 24$ giving $B = \pi/12$.

The temperature oscillates around an average value of 60° F, so $C = 60$. The amplitude of the oscillation is 20° F. To arrange that the temperature be at its lowest when $t = 0$, we take A negative, so $A = -20$. Thus

$$A = 60 - 20\cos\left(\frac{\pi}{12}t\right).$$

Solutions for Chapter 1 Review

1. The statement $f(9) = 14$ tells us that $W = 14$ when $t = 9$. In other words, in 1999, Argentina produced 14 million metric tons of wheat.

5. (a)

$$\begin{aligned}\text{Change between 1989 and 1997} &= \text{Revenues in 1997} - \text{Revenues in 1989}\\ &= 172{,}000 - 123{,}212\\ &= 48{,}788 \text{ million dollars.}\end{aligned}$$

(b)

$$\begin{aligned}\begin{array}{c}\text{Average rate of change}\\ \text{between 1989 and 1997}\end{array} &= \frac{\text{Change in revenues}}{\text{Change in time}}\\ &= \frac{\text{Revenues in 1997} - \text{Revenues in 1989}}{1997 - 1989}\\ &= \frac{172{,}000 - 123{,}212}{1997 - 1989}\\ &= \frac{48{,}788}{8}\\ &= 6098.5 \text{ million dollars per year.}\end{aligned}$$

This means that General Motors' revenues increased on average by 6098.5 million dollars per year between 1989 and 1997.

(c) From 1987 to 1997 there were two one-year time intervals during which the average rate of change in revenues was negative: -1191 million dollars per year between 1989 and 1990, and -4760 million dollars per year between 1995 and 1996.

9. The equation of the line is of the form $y = b + mx$ where m is the slope given by

$$\text{Slope} = \frac{\text{Rise}}{\text{Run}} = \frac{0}{2-0} = 0$$

so the line is $y = 0 \cdot x + b = b$. Since the line passes through the point $(0, 2)$ we have $b = 2$ and the equation of the line is $y = 2$.

13. (a) Since the price is decreasing at a constant absolute rate, the price of the product is a linear function of t. In t days, the product will cost $80 - 4t$ dollars.

(b) Since the price is decreasing at a constant relative rate, the price of the product is an exponential function of t. In t days, the product will cost $80(0.95)^t$.

17. We are looking for a linear function $y = f(x)$ that, given a time x in years, gives a value y in dollars for the value of the refrigerator. We know that when $x = 0$, that is, when the refrigerator is new, $y = 950$, and when $x = 7$, the refrigerator is worthless, so $y = 0$. Thus $(0, 950)$ and $(7, 0)$ are on the line that we are looking for. The slope is then given by

$$m = \frac{950}{-7}$$

It is negative, indicating that the value decreases as time passes. Having found the slope, we can take the point $(7, 0)$ and use the point-slope formula:

$$y - y_1 = m(x - x_1).$$

So,

$$\begin{aligned} y - 0 &= -\frac{950}{7}(x - 7)\\ y &= -\frac{950}{7}x + 950.\end{aligned}$$

21. This is a line with slope $-3/7$ and y-intercept 3, so a possible formula is

$$y = -\frac{3}{7}x + 3.$$

25. $z = 1 - \cos\theta$

29. We divide both sides by 25 and then use logarithms:

$$\begin{aligned} e^{3x} &= \frac{10}{25} = 0.4 \\ \ln(e^{3x}) &= \ln(0.4) \\ 3x &= \ln(0.4) \\ x &= \frac{\ln(0.4)}{3} = -0.305. \end{aligned}$$

33. **(a)** The slope of the line is

$$\text{Slope} = \frac{\Delta C}{\Delta t} = \frac{369.4 - 338.5}{2000 - 1980} = 1.545 \text{ ppm/year.}$$

The initial value (when $t = 0$) is 338.5, so the linear formula is

$$C = 338.5 + 1.545t.$$

The absolute annual rate of increase in carbon dioxide concentration is 1.545 ppm per year.

(b) If the function is exponential, we have $C = C_0a^t$. The initial value (when $t = 0$) is 338.5, so we have $C = 338.5a^t$. We use the fact that $C = 369.4$ when $t = 20$ to find a:

$$\begin{aligned} 369.4 &= 338.5a^{20} \\ 1.0913 &= a^{20} \\ a &= (1.0913)^{1/20} = 1.0044. \end{aligned}$$

The exponential formula is

$$C = 338.5(1.0044)^t.$$

The relative annual rate of increase in carbon dioxide concentration is 0.44% per year. (Alternately, if we used the form $C = C_0e^{kt}$, the formula would be $C = 338.5e^{0.0044t}$.)

37. **(a)** We have $f(g(x)) = f(\ln x) = (\ln x)^2 + 1$.

(b) We have $g(f(x)) = g(x^2 + 1) = \ln(x^2 + 1)$.

(c) We have $f(f(x)) = f(x^2 + 1) = (x^2 + 1)^2 + 1 = x^4 + 2x^2 + 2$.

41. This looks like the graph of $y = -x^2$ shifted up 2 units and to the left 3 units. One possible formula is $y = -(x+3)^2+2$. Other answers are possible. You can check your answer by graphing it using a calculator or computer.

45. In effect, your friend is offering to give you \$17,000 now in return for the \$19,000 lottery payment one year from now. Since $19{,}000/17{,}000 = 1.11764\cdots$, your friend is charging you 11.7% interest per year, compounded annually. You can expect to get more by taking out a loan as long as the interest rate is less than 11.7%. In particular, if you take out a loan, you have the first lottery check of \$19,000 plus the amount you can borrow to be paid back by a single payment of \$19,000 at the end of the year. At 8.25% interest, compounded annually, the present value of 19,000 one year from now is $19{,}000/(1.0825) = 17{,}551.96$. Therefore the amount you can borrow is the total of the first lottery payment and the loan amount, that is, $19{,}000 + 17{,}551.96 = 36{,}551.96$. So you do better by taking out a one-year loan at 8.25% per year, compounded annually, than by accepting your friend's offer.

49. **(a)** The period looks to be about 12 months. This means that the number of mumps cases oscillates and repeats itself approximately once a year.

$$\text{Amplitude} = \frac{\max - \min}{2} = \frac{11{,}000 - 2000}{2} = 4500 \text{ cases}$$

This means that the minimum and maximum number of cases of mumps are within $9000 (= 4500 \cdot 2)$ cases of each other.

(b) Assuming cyclical behavior, the number of cases in 30 months will be the same as the number of cases in 6 months which is about 2000 cases. (30 months equals 2 years and six months). The number of cases in 45 months equals the number of cases in 3 years and 9 months which assuming cyclical behavior is the same as the number of cases in 9 months. This is about 2000 as well.

Solutions to Problems on Fitting Formulas to Data

1. (a) See Figure 1.27. The regression line fits the data reasonably well.

(b) The slope is 0.734 trillion dollars per year. The model says that gross world product has been increasing at a rate of approximately 0.734 trillion dollars per year.

(c) The year 2005 corresponds to $t = 55$ and we have $G = 3.543 + 0.734(55) = 43.913$. Predicted gross world product in 2005 is 43.913 trillion dollars. The year 2020 corresponds to $t = 70$ and we have $G = 3.543 + 0.734(70) = 54.923$. Predicted gross world product in 2020 is 54.923 trillion dollars. We should have much more confidence in the 2005 prediction because it is closer to the time of the data values used.

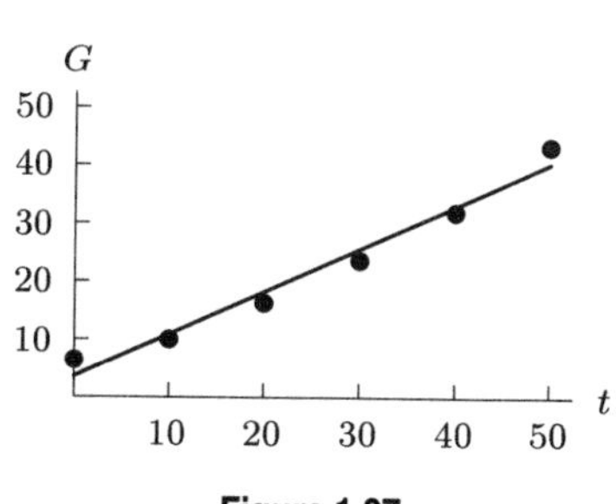

Figure 1.27

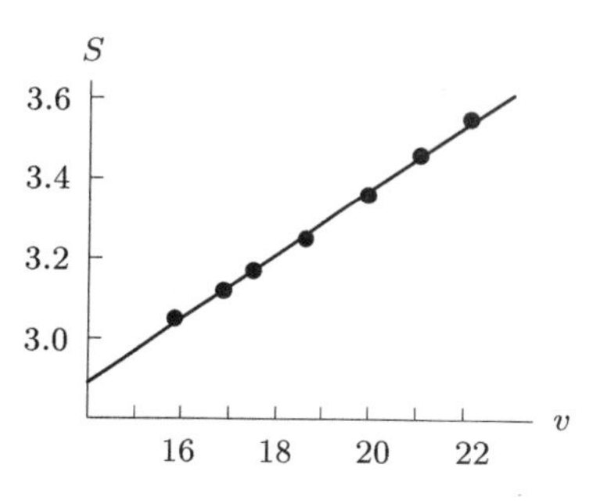

Figure 1.28

5. (a) Using a calculator, we get the stride rate S as a function of speed v,

$$S = 0.08v + 1.77.$$

(b) See Figure 1.28. The line seems to fit the data well.

(c) Using the formula, we substitute in the given speed of 18 ft/sec into our regression line and get

$$S = (0.08)(18) + 1.77 = 3.21.$$

To find the stride rate when the speed is 10 ft/sec, we substitute $v = 10$ and get

$$S = (0.08)(10) + 1.77 = 2.57.$$

We can have more confidence in our estimate for the stride at the speed 18 ft/sec rather than our estimate at 10 ft/sec, because the speed 18 ft/sec lies well within the domain of the data set given, and thus involves interpolation, whereas the speed 10 ft/sec lies well to the left of the plotted data points and is thus a more speculative value, involving extrapolation.

9. (a) The data is plotted in Figure 1.29.

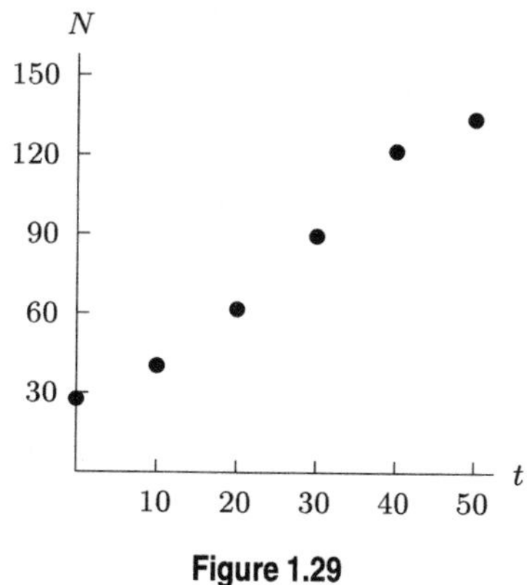

Figure 1.29

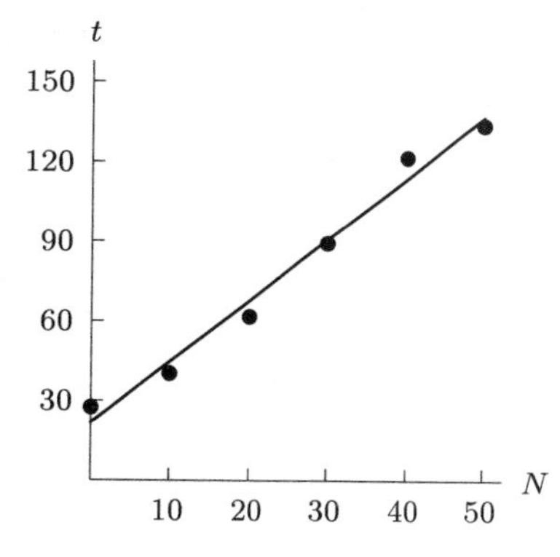

Figure 1.30

(b) It is not obvious which model is best.

(c) The regression line is

$$N = 2.3t + 21.7.$$

The line and the data are graphed in Figure 1.30. The estimated number of cars in the year 2010, according to this model is, substituting $t = 70$,

$$N = 2.3 \cdot 70 + 21.7 = 182.7 \text{ million}.$$

(d) The slope represents the fact that every year the number of passenger cars in the US grows by approximately 2,300,000.

(e) An exponential regression function that fits the data is

$$N = 29.7(1.034)^t.$$

(Different algorithms can give different formulas, so answers may vary.) The exponential curve and the data are graphed in Figure 1.31.

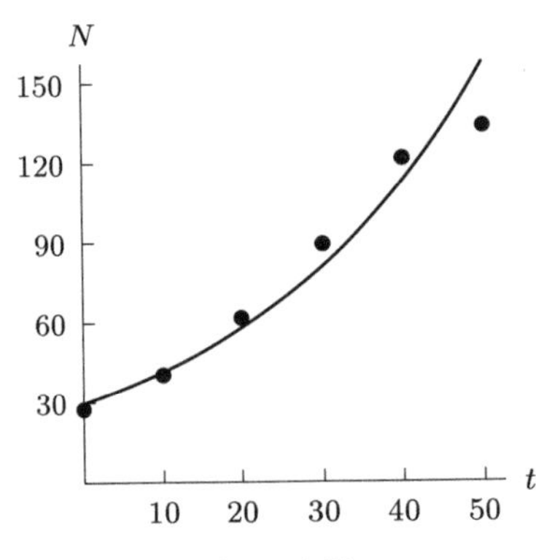

Figure 1.31

The estimated number of cars in the year 2010, according to this model is

$$N = 29.7(1.034)^{70} = 308.5 \text{ million}.$$

This is much larger than the prediction obtained from the linear model.

(f) The annual growth rate is as $a - 1$ for the exponential function $N = N_0 a^t$. Thus, the rate is 0.034 so the annual rate of growth in the number of US passenger cars is 3.4% according to our exponential model.

13. There are several possible answers to this questions depending on whether or not you choose to "smooth out" the given curve. In all cases the leading coefficient is positive. If you didn't "smooth out" the function, the graph changes direction 10 times so the degree of the polynomial would be at least 11. If you "smoothed" the function so that it changes direction 4 times, the function would have at least degree 5. Finally, if you "smoothed out" the function so that it only changes direction twice, the degree would be at least 3.

17. **(a)** Figure 1.32 shows the data with the logarithmic curve from part (c).

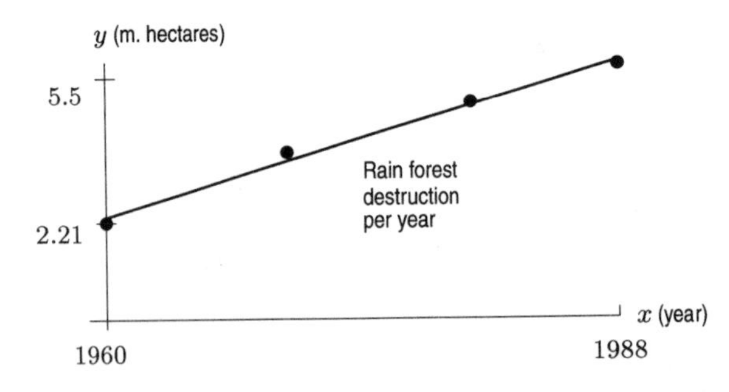

Figure 1.32

(b) The data are increasing and the curve is slightly concave down. The fact that the data are increasing means that more hectares of rain forest are being destroyed every year. The fact that the curve is concave down means the rate of increase of rain forest destruction is decreasing.

(c) Using a graphing calculator, we get $y \approx -1884.66 + 248.92 \ln x$. (Since different algorithms can give different formulas, answers may vary.)

(d) When $x = 2010$, we have $y \approx -1884.66 + 248.92 \ln 2010 = 8.6$ million hectares.

Solutions to Problems on Compound Interest and the Number e

1. The interest charged on the first month's debt incurs additional interest in the second and subsequent months. The total interest is greater than $12 \times 2\% = 24\%$.

At the end of 1 year, a debt of \$1 becomes a debt of $(1.02)^{12} \approx \$1.27$, so the annual percentage rate is approximately 27%.

5. In one year, an investment of P_0 becomes $P_0e^{0.06}$. Using a calculator, we see that

$$P_0e^{0.06} = P_0(1.0618365)$$

So the effective annual yield is about 6.18%.

9. **(a)** Substitute $n = 10{,}000$, then $n = 100{,}000$ and $n = 1{,}000{,}000$.

$$\left(1 + \frac{0.04}{10{,}000}\right)^{10{,}000} \approx 1.0408107$$

$$\left(1 + \frac{0.04}{100{,}000}\right)^{100{,}000} \approx 1.0408108$$

$$\left(1 + \frac{0.04}{1{,}000{,}000}\right)^{1{,}000{,}000} \approx 1.0408108$$

Effective annual yield:
4.08108%

(b) Evaluating gives $e^{0.04} \approx 1.048108$ as expected.

13. **(a)** Its cost in 1989 would be $1000 + \frac{1290}{100} \cdot 1000 = 13{,}900$ cruzados.

(b) The monthly inflation rate r solves

$$\left(1 + \frac{1290}{100}\right)^1 = \left(1 + \frac{r}{100}\right)^{12},$$

since we compound the monthly inflation 12 times to get the yearly inflation. Solving for r, we get $r = 24.52\%$. Notice that this is much different than $\frac{1290}{12} = 107.5\%$.

Solutions to Problems on Limits to Infinity and End Behavior

1. As $x \to \infty$, a power function with a larger exponent will dominate a power function with a smaller exponent. The function with the largest values is $y = 0.1x^4$ and the function with the smallest values is $y = 1000x^2$. A global picture is shown in Figure 1.33.

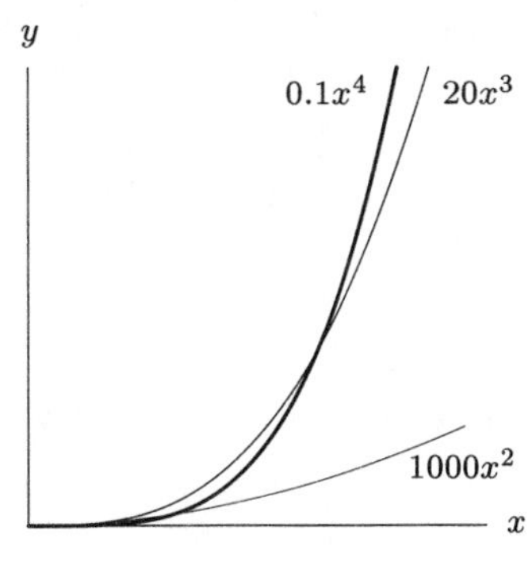

Figure 1.33

5. As $x \to \infty$, $f(x) = x^5$ has the largest positive values. As $x \to -\infty$, $g(x) = -x^3$ has the largest positive values.

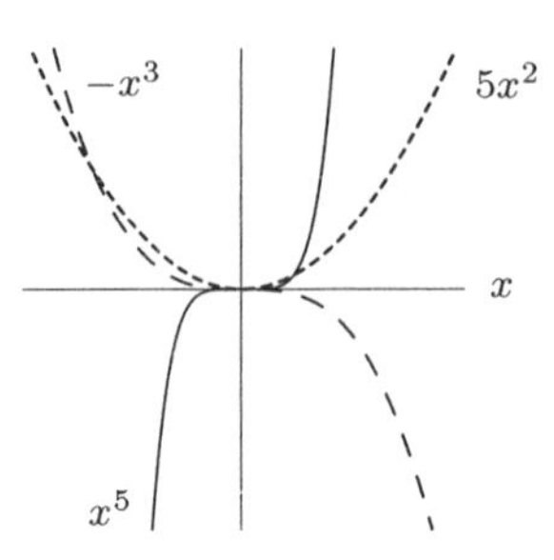

9. A possible graph is shown in Figure 1.34. There are many possible answers.

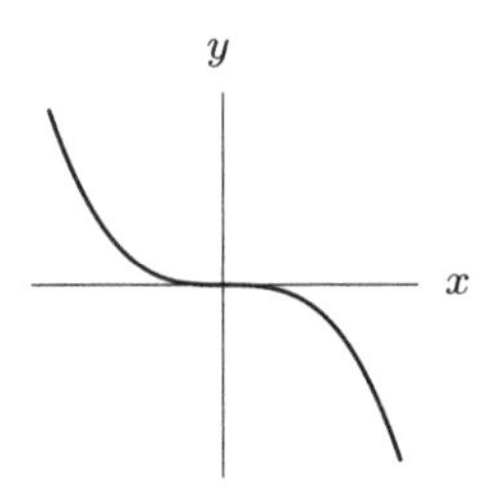

Figure 1.34

13. The values in Table 1.2 suggest that this limit is 0. The graph of $y = 1/x$ in Figure 1.35 suggests that $y \to 0$ as $x \to \infty$ and so supports the conclusion.

Table 1.2

x	100	1000	1,000,000
$1/x$	0.01	0.001	0.000001

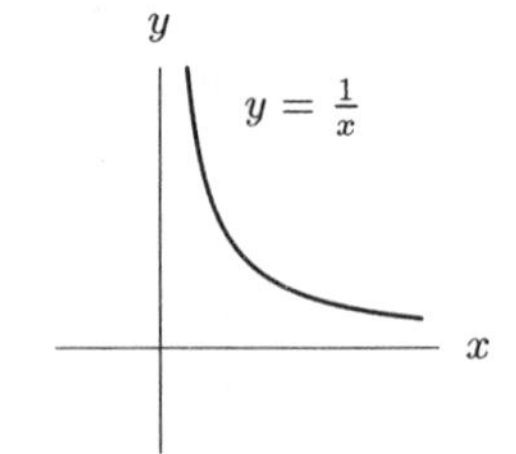

Figure 1.35

17. We see from a graph of $f(x) = 8(1 - e^{-x})$ that $\lim_{x\to\infty} f(x) = 8$ and $\lim_{x\to-\infty} f(x) = -\infty$.

21. A power function with a positive exponent always grows faster than $\ln x$, so the answer is $x^{1/2}$.

25. As $x \to \infty$, a power function with a larger exponent will dominate a power function with a smaller exponent. We see that A corresponds to $y = 0.2x^5$, B corresponds to $y = x^4$, C corresponds to $y = 5x^3$, and D corresponds to $y = 70x^2$.

29.

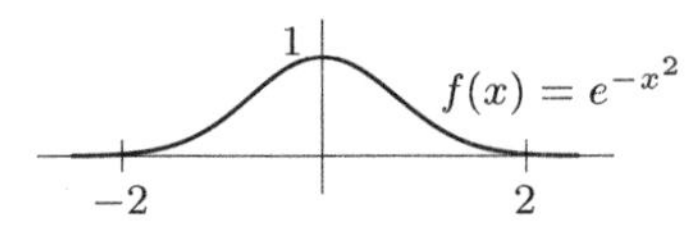

Figure 1.36

(a) The function f is increasing for negative x. The function f is decreasing for all positive x.
(b) The graph of f is concave down near $x = 0$.
(c) As $x \to \infty$, $f(x) \to 0$. As $x \to -\infty$, $f(x) \to 0$.

CHAPTER TWO

Solutions for Section 2.1

1. **(a)** The average velocity between $t = 3$ and $t = 5$ is

$$\frac{\text{Distance}}{\text{Time}} = \frac{s(5) - s(3)}{5 - 3} = \frac{25 - 9}{2} = \frac{16}{2} = 8 \text{ ft/sec.}$$

(b) Using an interval of size 0.1, we have

$$\begin{pmatrix} \text{Instantaneous velocity} \\ \text{at } t = 3 \end{pmatrix} \approx \frac{s(3.1) - s(3)}{3.1 - 3} = \frac{9.61 - 9}{0.1} = 6.1.$$

Using an interval of size 0.01, we have

$$\begin{pmatrix} \text{Instantaneous velocity} \\ \text{at } t = 3 \end{pmatrix} \approx \frac{s(3.01) - s(3)}{3.01 - 3} = \frac{9.0601 - 9}{0.01} = 6.01.$$

From this we guess that the instantaneous velocity at $t = 3$ is about 6 ft/sec.

5.

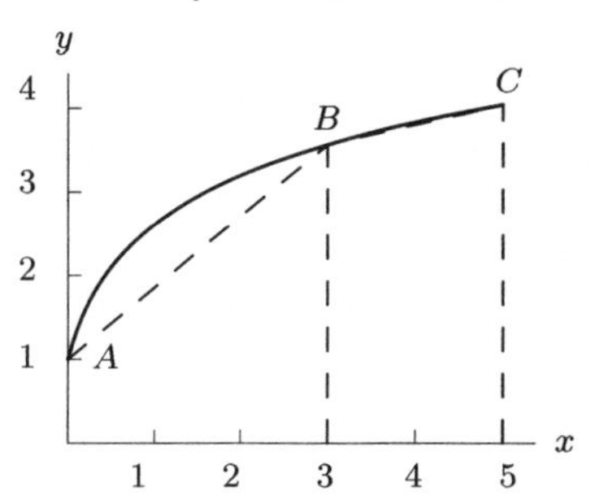

Figure 2.1

(a) The average rate of change of a function over an interval is represented graphically as the slope of the secant line to its graph over the interval. Segment AB is the secant line to the graph in the interval from $x = 0$ to $x = 3$ and segment BC is the secant line to the graph in the interval from $x = 3$ to $x = 5$.

We can easily see that slope of $AB >$ slope of BC. Therefore, the average rate of change between $x = 0$ and $x = 3$ is greater than the average rate of change between $x = 3$ and $x = 5$.

(b) We can see from the graph in Figure 2.2 that the function is increasing faster at $x = 1$ than at $x = 4$. Therefore, the instantaneous rate of change at $x = 1$ is greater than the instantaneous rate of change at $x = 4$.

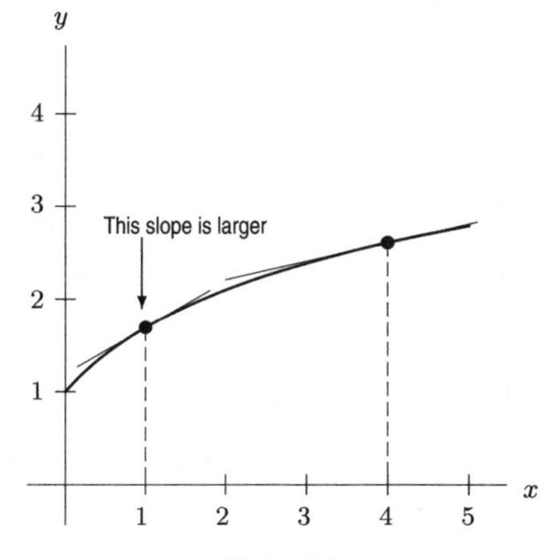

Figure 2.2

(c) The units of rate of change are obtained by dividing units of cost by units of product: thousands of dollars/kilogram.

9. We estimate $f'(2)$ using the average rate of change formula on a small interval around 2. We use the interval $x = 2$ to $x = 2.001$. (Any small interval around 2 gives a reasonable answer.) We have

$$f'(2) \approx \frac{f(2.001) - f(2)}{2.001 - 2} = \frac{3^{2.001} - 3^2}{2.001 - 2} = \frac{9.00989 - 9}{0.001} = 9.89.$$

13. Since $f'(x) = 0$ where the graph is horizontal, $f'(x) = 0$ at $x = d$. The derivative is positive at points b and c, but the graph is steeper at $x = c$. Thus $f'(x) = 0.5$ at $x = b$ and $f'(x) = 2$ at $x = c$. Finally, the derivative is negative at points a and e but the graph is steeper at $x = e$. Thus, $f'(x) = -0.5$ at $x = a$ and $f'(x) = -2$ at $x = e$. See Table 2.1.

Thus, we have $f'(d) = 0$, $f'(b) = 0.5$, $f'(c) = 2$, $f'(a) = -0.5$, $f'(e) = -2$.

Table 2.1

x	$f'(x)$
d	0
b	0.5
c	2
a	-0.5
e	-2

17. **(a)** The function $N = f(t)$ is decreasing when $t = 1950$. Therefore, $f'(1950)$ is negative. That means that the number of farms in the United States, in millions, was decreasing in 1950.

(b) The function $N = f(t)$ is decreasing in 1960 as well as in 1980 but it is decreasing faster in 1960 than in 1980. Therefore, $f'(1960)$ is more negative than $f'(1980)$.

21. The coordinates of A are $(4, 25)$. See Figure 2.3. The coordinates of B and C are obtained using the slope of the tangent line. Since $f'(4) = 1.5$, the slope is 1.5

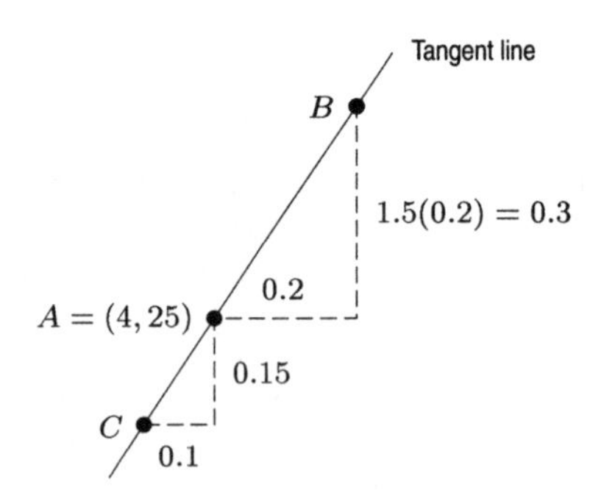

Figure 2.3

From A to B, $\Delta x = 0.2$, so $\Delta y = 1.5(0.2) = 0.3$. Thus, at C we have $y = 25 + 0.3 = 25.3$. The coordinates of B are $(4.2, 25.3)$.

From A to C, $\Delta x = -0.1$, so $\Delta y = 1.5(-0.1) = -0.15$. Thus, at C we have $y = 25 - 0.15 = 24.85$. The coordinates of C are $(3.9, 24.85)$.

25. Using a difference quotient with $h = 0.001$, say, we find

$$f'(1) \approx \frac{1.001\ln(1.001) - 1\ln(1)}{1.001 - 1} = 1.0005$$

$$f'(2) \approx \frac{2.001\ln(2.001) - 2\ln(2)}{2.001 - 2} = 1.6934$$

The fact that f' is larger at $x = 2$ than at $x = 1$ suggests that f is concave up on the interval $[1, 2]$.

Solutions for Section 2.2

1. The function is decreasing for $x < -2$ and $x > 2$, and increasing for $-2 < x < 2$. The matching derivative must be negative (below the x-axis) for $x < -2$ and $x > 2$, positive (above the x-axis) for $-2 < x < 2$, and zero (on the x-axis) for $x = -2$ and $x = 2$. The matching derivative is in graph VIII.

5. **(a)** We use the interval to the right of $x = 2$ to estimate the derivative. (Alternately, we could use the interval to the left of 2, or we could use both and average the results.) We have

$$f'(2) \approx \frac{f(4) - f(2)}{4 - 2} = \frac{24 - 18}{4 - 2} = \frac{6}{2} = 3.$$

We estimate $f'(2) \approx 3$.

(b) We know that $f'(x)$ is positive when $f(x)$ is increasing and negative when $f(x)$ is decreasing, so it appears that $f'(x)$ is positive for $0 < x < 4$ and is negative for $4 < x < 12$.

9.

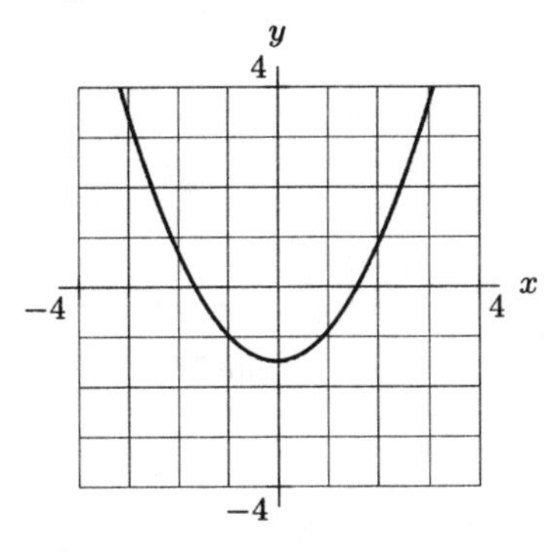

13. Since $f'(x) > 0$ for $x < -1$, $f(x)$ is increasing on this interval.
Since $f'(x) < 0$ for $x > -1$, $f(x)$ is decreasing on this interval.
Since $f'(x) = 0$ at $x = -1$, the tangent to $f(x)$ is horizontal at $x = -1$.
One possible shape for $y = f(x)$ is shown in Figure 2.4.

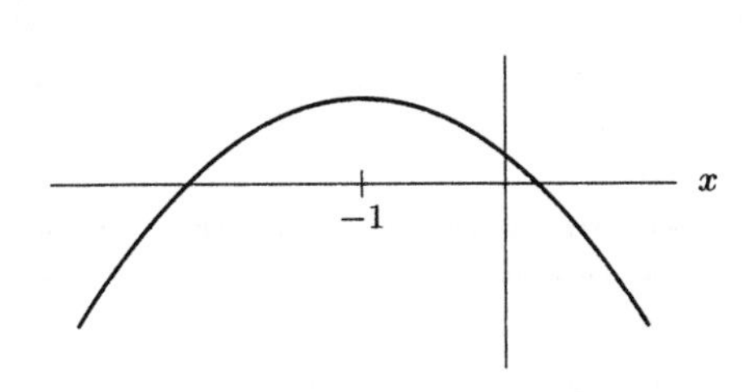

Figure 2.4

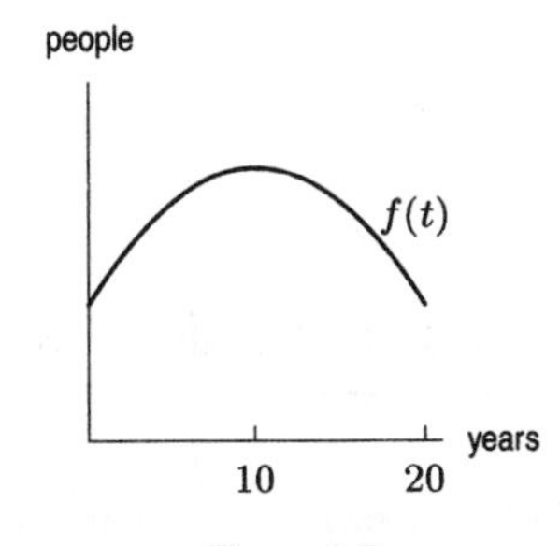

Figure 2.5

17. The graph is increasing for $0 < t < 10$ and is decreasing for $10 < t < 20$. One possible graph is shown in Figure 2.5. The units on the horizontal axis are years and the units on the vertical axis are people.

The derivative is positive for $0 < t < 10$ and negative for $10 < t < 20$. Two possible graphs are shown in Figure 2.6. The units on the horizontal axes are years and the units on the vertical axes are people per year.

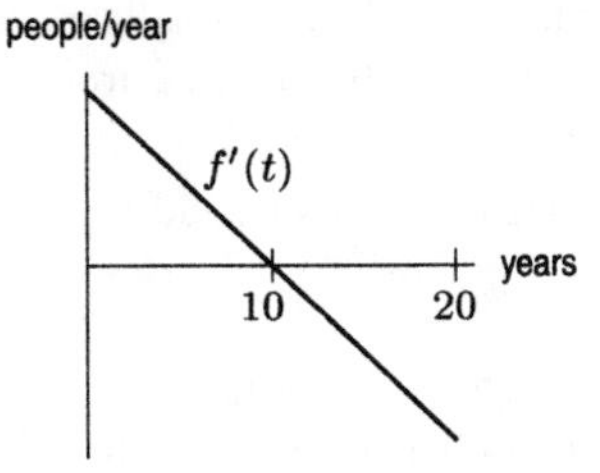

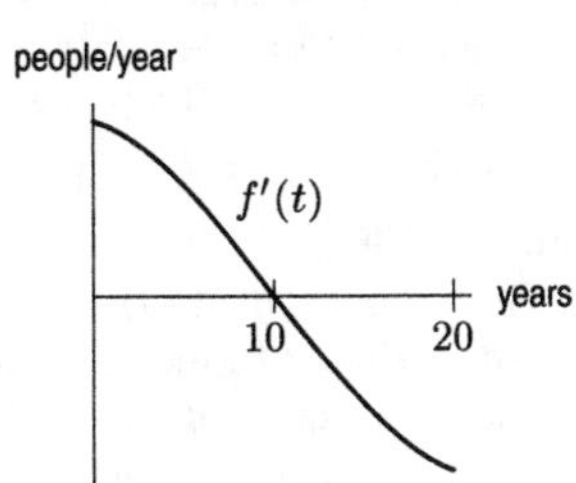

Figure 2.6

21. The function is increasing for $x < 1$ and decreasing for $x > 1$, so the derivative is positive for $x < 1$ and negative for $x > 1$. One possible graph is in Figure 2.7.

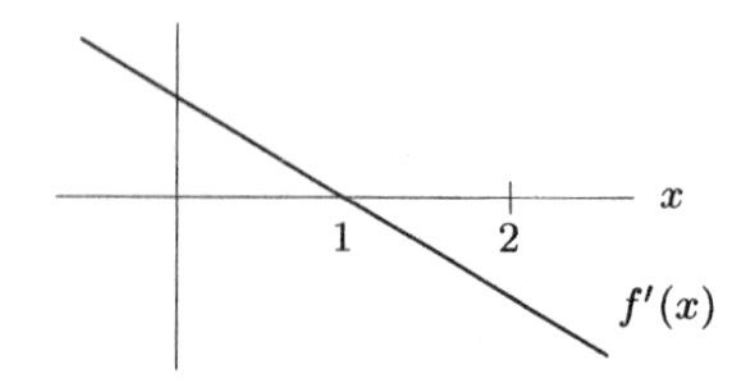

Figure 2.7

25. The function is increasing and concave up, so $f'(x)$ is positive and increasing. See Figure 2.8.

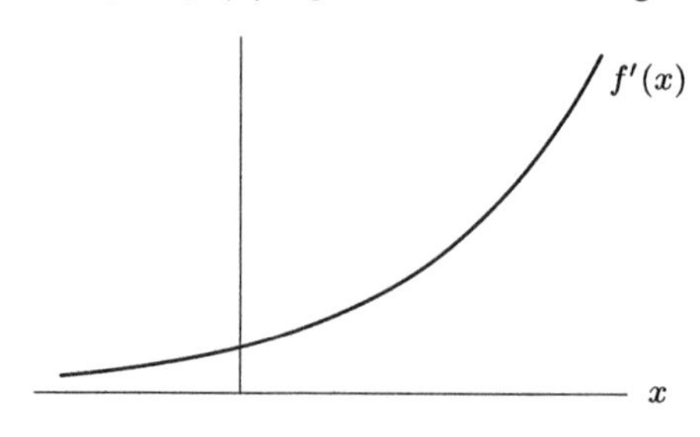

Figure 2.8

29. **(a)** $t = 3$
(b) $t = 9$
(c) $t = 14$
(d)

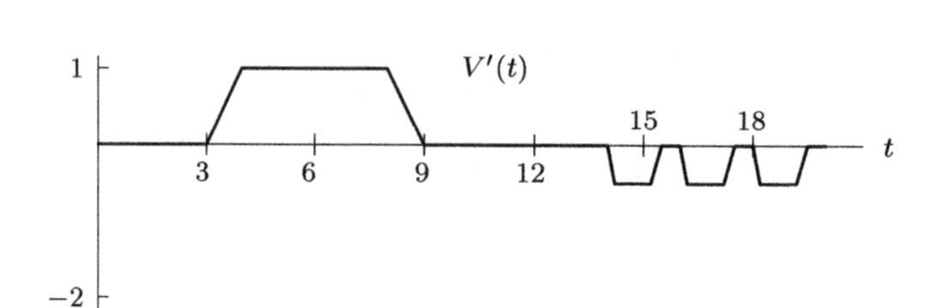

Solutions for Section 2.3

1. **(a)** As the cup of coffee cools, the temperature decreases, so $f'(t)$ is negative.
(b) Since $f'(t) = dH/dt$, the units are degrees Celsius per minute. The quantity $f'(20)$ represents the rate at which the coffee is cooling, in degrees per minute, 20 minutes after the cup is put on the counter.

5. **(a)** Since $f'(c)$ is negative, the function $P = f(c)$ is decreasing: pelican eggshells are getting thinner as the concentration, c, of PCBs in the environment is increasing.
(b) The statement $f(200) = 0.28$ means that the thickness of pelican eggshells is 0.28 mm when the concentration of PCBs in the environment is 200 parts per million (ppm).

The statement $f'(200) = -0.0005$ means that the thickness of pelican eggshells is decreasing (eggshells are becoming thinner) at a rate of 0.0005 mm per ppm of concentration of PCBs in the environment when the concentration of PCBs in the environment is 200 ppm.

9. **(a)** The statement $f(140) = 120$ means that a patient weighing 140 pounds should receive a dose of 120 mg of the painkiller. The statement $f'(140) = 3$ tells us that if the weight of a patient increases by about one pound (from 140 pounds), the dose should be increased by about 3 mg.
(b) Since the dose for a weight of 140 lbs is 120 mg and at this weight the dose goes up by 3 mg for each pound, a 145 lb patient should get an additional $3(5) = 15$ mg. Thus, for a 145 lb patient, the correct dose is approximately

$$f(145) \approx 120 + 3(5) = 135 \text{ mg}.$$

13. **(a)** Since t represents the number of days from now, we are told $f(0) = 80$ and $f'(0) = 0.50$.
(b)

$$\begin{aligned} f(10) &\approx \text{value now} + \text{change in value in 10 days} \\ &= 80 + 0.50(10) \\ &= 80 + 5 \\ &= 85. \end{aligned}$$

In 10 days, we expect that the mutual fund will be worth about \$85 a share.

17. **(a)** This means that investing the \$1000 at 5% would yield \$1649 after 10 years.
(b) Writing $g'(r)$ as dB/dt, we see that the units of dB/dt are dollars per percent (interest). We can interpret dB as the extra money earned if interest rate is increased by dr percent. Therefore $g'(5) = \frac{dB}{dr}|_{r=5} \approx 165$ means that the balance, at 5% interest, would increase by about \$165 if the interest rate were increased by 1%. In other words, $g(6) \approx g(5) + 165 = 1649 + 165 = 1814$.

21. **(a)** If the price is \$150, then 2000 items will be sold.
(b) If the price goes up from \$150 by \$1 per item, about 25 fewer items will be sold. Equivalently, if the price is decreased from \$150 by \$1 per item, about 25 more items will be sold.

Solutions for Section 2.4

1. **(a)** Since the graph is below the x-axis at $x = 2$, the value of $f(2)$ is negative.
(b) Since $f(x)$ is decreasing at $x = 2$, the value of $f'(2)$ is negative.
(c) Since $f(x)$ is concave up at $x = 2$, the value of $f''(2)$ is positive.

5. The graph must be everywhere decreasing and concave up on some intervals and concave down on other intervals. One possibility is shown in Figure 2.9.

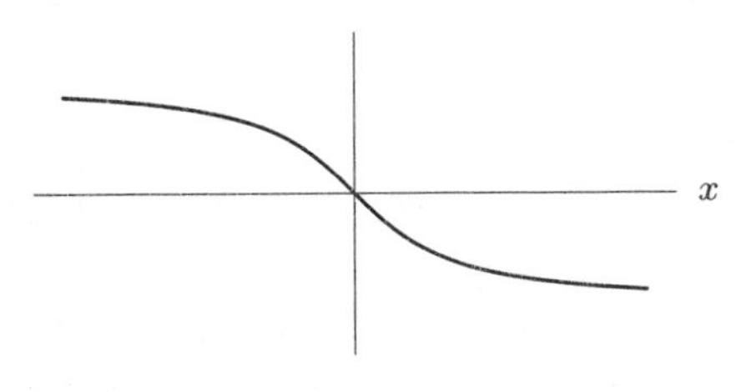

Figure 2.9

9. $f'(x) < 0$
$f''(x) = 0$

13. **(a)** The derivative, $f'(t)$, appears to be positive since the number of cars is increasing. The second derivative, $f''(t)$, appears to be positive because the rate of change is increasing. For example, between 1940 and 1950, the rate of change is $(40.3 - 27.5)/10 = 1.28$ million cars per year, while between 1950 and 1960, the rate of change is 2.14 million cars per year.
(b) We use the average rate of change formula on the interval 1970 to 1980 to estimate $f'(1975)$:

$$f'(1975) \approx \frac{121.6 - 89.3}{1980 - 1970} = \frac{32.3}{10} = 3.23.$$

We see that $f'(1975) \approx 3.23$ million cars per year. The number of passenger cars in the US was increasing at a rate of about 3.23 million cars per year in 1975.

17. The derivative of $w(t)$ appears to be negative since the function is decreasing over the interval given. The second derivative, however, appears to be positive since the function is concave up, i.e., it is decreasing at a decreasing rate.

21. Since all advertising campaigns are assumed to produce an increase in sales, a graph of sales against time would be expected to have a positive slope.

A positive second derivative means the rate at which sales are increasing is increasing. If a positive second derivative is observed during a new campaign, it is reasonable to conclude that this increase in the rate sales are increasing is caused by the new campaign–which is therefore judged a success. A negative second derivative means a decrease in the rate at which sales are increasing, and therefore suggests the new campaign is a failure.

25. **(a)** For 1989 we have

$$\frac{dP}{dt} \approx \frac{54.1 - 62.4}{1989 - 1986} = \frac{-8.3}{3} \approx -2.77 \text{ \%/year}.$$

For 1992 we have

$$\frac{dP}{dt} \approx \frac{48.0 - 54.1}{1992 - 1989} = \frac{-6.1}{3} \approx -2.03 \text{ \%/year}.$$

For 1995 we have

$$\frac{dP}{dt} \approx \frac{43.5 - 48.0}{1995 - 1992} = \frac{-4.5}{3} \approx -1.50 \text{ \%/year}.$$

For 1998 we have

$$\frac{dP}{dt} \approx \frac{41.8 - 43.5}{1998 - 1995} = \frac{-1.7}{3} \approx -0.57 \text{ \%/year}.$$

(b) Since $\frac{dP}{dt}$ is increasing from 1989 to 2001 we get that $\frac{d^2P}{dt^2}$ is positive.

(c) The values of P and $\frac{dP}{dt}$ are troublesome because they indicate that the percent of students graduating is low, and that the number is getting smaller each year.

(d) Since $\frac{d^2P}{dt^2}$ is positive, the percent of students graduating is not decreasing as fast as it once was. Also, in 1998 dP/dt has a magnitude of less than 1% a year so the level of drop-outs does in fact seem to be hitting its minimum at around 40%.

Solutions for Section 2.5

1. We know $MC \approx C(1{,}001) - C(1{,}000)$. Therefore, $C(1{,}001) \approx C(1{,}000) + MC$ or $C(1{,}001) \approx 5000 + 25 = 5025$ dollars.

Since we do not know $MC(999)$, we will assume that $MC(999) = MC(1{,}000)$. Therefore:

$$MC(999) = C(1{,}000) - C(999).$$

Then:

$$C(999) \approx C(1{,}000) - MC(999) = 5{,}000 - 25 = 4{,}975 \text{ dollars}.$$

Alternatively, we can reason that

$$MC(1{,}000) \approx C(1{,}000) - C(999),$$

so

$$C(999) \approx C(1{,}000) - MC(1{,}000) = 4{,}975 \text{ dollars}.$$

Now for $C(1{,}000)$, we have

$$C(1{,}100) \approx C(1{,}000) + MC \cdot 100.$$

Since $1{,}100 - 1{,}000 = 100$,

$$C(1{,}100) \approx 5{,}000 + 25 \times 100 = 5{,}000 + 2{,}500 = 7{,}500 \text{ dollars}.$$

5. Marginal cost $= C'(q)$. Therefore, marginal cost at q is the slope of the graph of $C(q)$ at q. We can see that the slope at $q = 5$ is greater than the slope at $q = 30$. Therefore, marginal cost is greater at $q = 5$. At $q = 20$, the slope is small, whereas at $q = 40$ the slope is larger. Therefore, marginal cost at $q = 40$ is greater than marginal cost at $q = 20$.

9. **(a)** The value of $C(0)$ represents the fixed costs before production, that is, the cost of producing zero units, incurred for initial investments in equipment, and so on.

(b) The marginal cost decreases slowly, and then increases as quantity produced increases. See Problem 7, graph (b).

(c) Concave down implies decreasing marginal cost, while concave up implies increasing marginal cost.

(d) An inflection point of the cost function is (locally) the point of maximum or minimum marginal cost.

(e) One would think that the more of an item you produce, the less it would cost to produce extra items. In economic terms, one would expect the marginal cost of production to decrease, so we would expect the cost curve to be concave down. In practice, though, it eventually becomes more expensive to produce more items, because workers and resources may become scarce as you increase production. Hence after a certain point, the marginal cost may rise again. This happens in oil production, for example.

13. **(a)** The profit earned by the 51st is the revenue earned by the 51st item minus the cost of producing the 51st item. This can be approximated by

$$\pi'(50) = R'(50) - C'(50) = 84 - 75 = \$9.$$

Thus the profit earned from the 51st item will be approximately \$9.

(b) The profit earned by the 91st item will be the revenue earned by the 91st item minus the cost of producing the 91st item. This can be approximated by

$$\pi'(90) = R'(90) - C'(90) = 68 - 71 = -\$3.$$

Thus, approximately three dollars are lost in the production of the 91st item.

(c) If $R'(78) > C'(78)$, production of a 79th item would increase profit. If $R'(78) < C'(78)$, production of one less item would increase profit. Since profit is maximized at $q = 78$, we must have

$$C'(78) = R'(78).$$

Solutions for Chapter 2 Review

1.

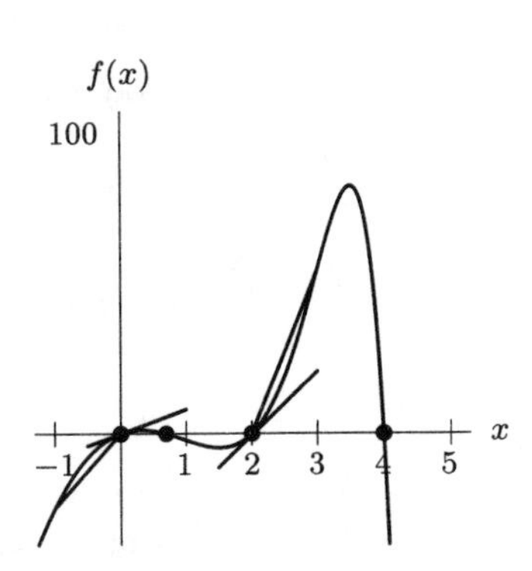

(a) The graph shows four different zeros in the interval, at $x = 0$, $x = 2$, $x = 4$ and $x \approx 0.7$.

(b) At $x = 0$ and $x = 2$, we see that the tangent has a positive slope so f is increasing.
At $x = 4$, we notice that the tangent to the curve has negative slope, so f is decreasing.

(c) Comparing the slopes of the secant lines at these values, we can see that the average rate of change of f is greater on the interval $2 \leq x \leq 3$.

(d) Looking at the tangents of the function at $x = 0$ and $x = 2$, we see that the slope of the tangent at $x = 2$ is greater. Thus, the instantaneous rate of change of f is greater at $x = 2$.

5. Using the interval $1 \leq x \leq 1.001$, we estimate

$$f'(1) \approx \frac{f(1.001) - f(1)}{0.001} = \frac{3.0033 - 3.0000}{0.001} = 3.3$$

The graph of $f(x) = 3^x$ is concave up so we expect our estimate to be greater than $f'(1)$.

9.

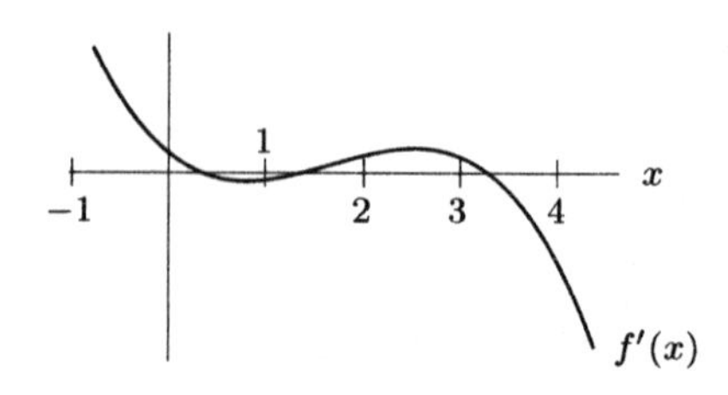

13. Using $\Delta x = 1$, we can say that

$$\begin{aligned} f(21) &= f(20) + \text{change in } f(x) \\ &\approx f(20) + f'(20)\Delta x \\ &= 68 + (-3)(1) \\ &= 65. \end{aligned}$$

Similarly, using $\Delta x = -1$,

$$\begin{aligned} f(19) &= f(20) + \text{change in } f(x) \\ &\approx f(20) + f'(20)\Delta x \\ &= 68 + (-3)(-1) \\ &= 71. \end{aligned}$$

Using $\Delta x = 5$, we can write

$$\begin{aligned} f(25) &= f(20) + \text{change in } f(x) \\ &\approx f(20) + (-3)(5) \\ &= 68 - 15 \\ &= 53. \end{aligned}$$

17.

(a)

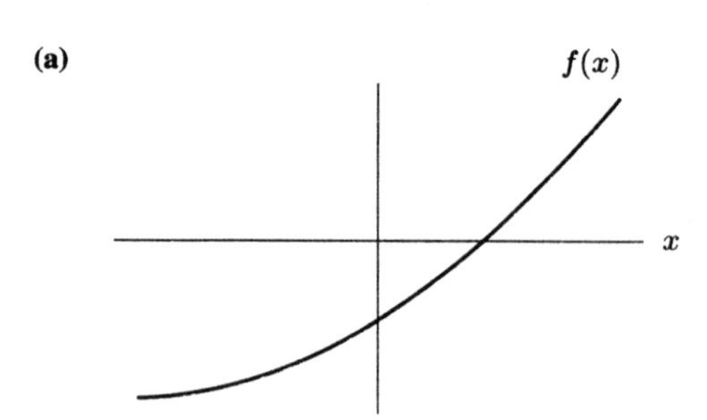

(b)

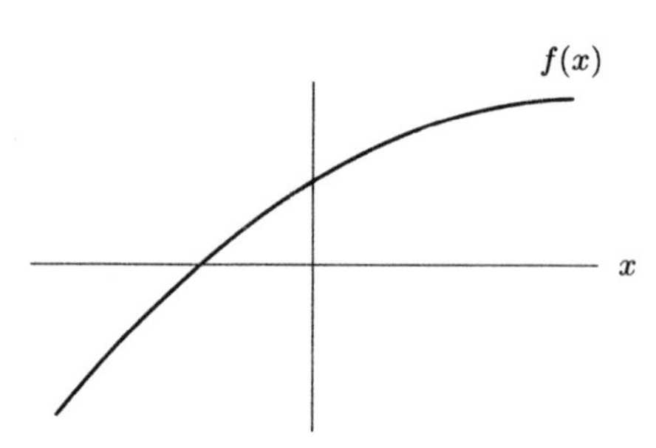

(c)

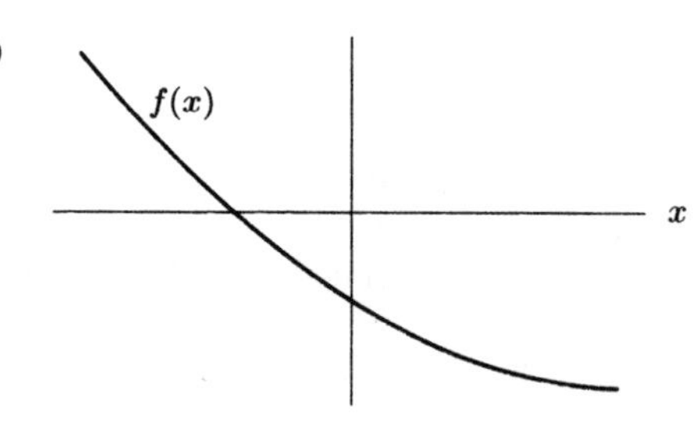

(d)

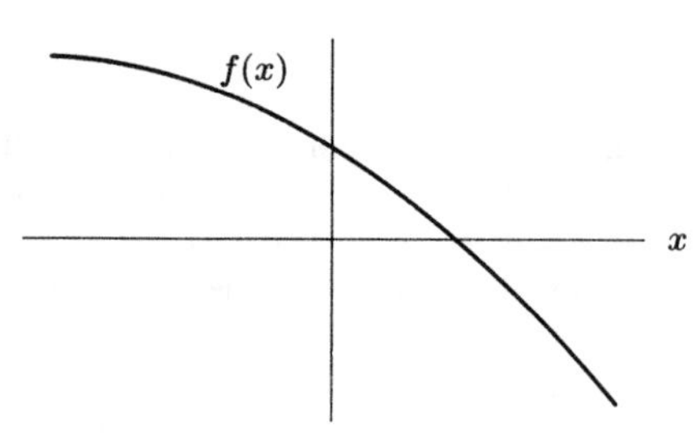

21. **(a)** Since the point $A = (7, 3)$ is on the graph of f, we have $f(7) = 3$.
(b) The slope of the tangent line touching the curve at $x = 7$ is given by

$$\text{Slope} = \frac{\text{Rise}}{\text{Run}} = \frac{3.8 - 3}{7.2 - 7} = \frac{0.8}{0.2} = 4.$$

Thus, $f'(7) = 4$.

25. **(a)** Since the traffic flow is the number of cars per hour, it is the slope of the graph of $C(t)$. It is greatest where the graph of the function $C(t)$ is the steepest and increasing. This happens at approximately $t = 3$ hours, or 7am.

(b) By reading the values of $C(t)$ from the graph we see:

$$\begin{array}{c}\text{Average rate of change}\\ \text{on } 1 \le t \le 2\end{array} = \frac{C(2) - C(1)}{2 - 1} = \frac{1000 - 400}{1} = \frac{600}{1} = 600 \text{ cars/hour},$$

$$\begin{array}{c}\text{Average rate of change}\\ \text{on } 2 \le t \le 3\end{array} = \frac{C(3) - C(2)}{3 - 2} = \frac{2000 - 1000}{1} = \frac{1000}{1} = 1000 \text{ cars/hour}.$$

A good estimate of $C'(2)$ is the average of the last two results. Therefore:

$$C'(2) \approx \frac{(600 + 1000)}{2} = \frac{1600}{2} = 800 \text{ cars/hour}.$$

(c) Since $t = 2$ is 6 am, the fact that $C'(2) \approx 800$ cars/hour means that the traffic flow at 6 am is about 800 cars/hour.

29. **(a)** The statement $f(15) = 200$ tells us that when the price is \$15, we sell about 200 units of the product.

(b) The statement $f'(15) = -25$ tells us that if we increase the price by \$1 (from 15), we will sell about 25 fewer units of the product.

Solutions to Problems on Limits and the Definition of the Derivative

1. The answers to parts (a)–(f) are marked in Figure 2.10.

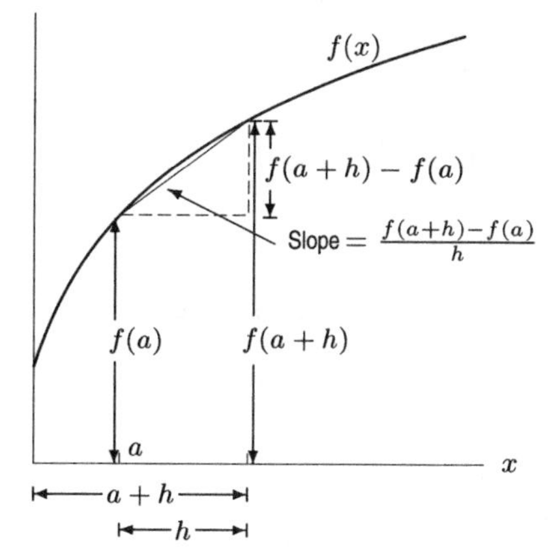

Figure 2.10

5. Using $h = 0.1, 0.01, 0.001$, we see

$$\frac{(3 + 0.1)^3 - 27}{0.1} = 27.91$$
$$\frac{(3 + 0.01)^3 - 27}{0.01} = 27.09$$
$$\frac{(3 + 0.001)^3 - 27}{0.001} = 27.009.$$

These calculations suggest that $\lim_{h \to 0} \frac{(3 + h)^3 - 27}{h} = 27$.

9. Yes, $f(x)$ is continuous on $0 \le x \le 2$.

13. Yes: $f(x) = x + 2$ is continuous for all values of x.

17. No: $f(x) = 1/(x - 1)$ is not continuous on any interval containing $x = 1$.

21. Since we can't make a fraction of a pair of pants, the number increases in jumps, so the function is not continuous.

25. Using the definition of the derivative, we have

$$\begin{aligned} f'(x) &= \lim_{h\to 0} \frac{f(x+h)-f(x)}{h} \\ &= \lim_{h\to 0} \frac{(3(x+h)-2)-(3x-2)}{h} \\ &= \lim_{h\to 0} \frac{3x+3h-2-3x+2}{h} \\ &= \lim_{h\to 0} \frac{3h}{h}. \end{aligned}$$

As h gets very close to zero without actually equaling zero, we can cancel the h in the numerator and denominator to obtain

$$f'(x) = \lim_{h\to 0}(3) = 3.$$

29. Using the definition of the derivative, we have

$$\begin{aligned} f'(x) &= \lim_{h\to 0} \frac{f(x+h)-f(x)}{h} \\ &= \lim_{h\to 0} \frac{(2(x+h)^2+(x+h))-(2x^2+x)}{h} \\ &= \lim_{h\to 0} \frac{2(x^2+2xh+h^2)+(x+h)-2x^2-x}{h} \\ &= \lim_{h\to 0} \frac{2x^2+4xh+2h^2+x+h-2x^2-x}{h} \\ &= \lim_{h\to 0} \frac{4xh+2h^2+h}{h} \\ &= \lim_{h\to 0} \frac{h(4x+2h+1)}{h}. \end{aligned}$$

As h gets very close to zero without actually equaling zero, we can cancel the h in the numerator and denominator to obtain

$$f'(x) = \lim_{h\to 0}(4x+2h+1) = 4x+1.$$

CHAPTER THREE

Solutions for Section 3.1

1. $\dfrac{dy}{dx} = 0$
5. $y' = -12x^{-13}$.
9. $f'(q) = 3q^2$
13. $\dfrac{dy}{dt} = 24t^2 - 8t + 12$.
17. $y' = 2z - \frac{1}{2z^2}$.
21. $y' = 2ax + b$.
25. Since $4/3$, π, and b are all constants, we have

$$\frac{dV}{dr} = \frac{4}{3}\pi(2r)b = \frac{8}{3}\pi rb.$$

29. $f'(x) = 3x^2 - 8x + 7$, so $f'(0) = 7$, $f'(2) = 3$, and $f'(-1) = 18$.
33. Since $f(t) = 700 - 3t^2$, we have $f(5) = 700 - 3(25) = 625$ cm. Since $f'(t) = -6t$, we have $f'(5) = -30$ cm/year. In the year 2000, the sand dune was 625 cm high and it was eroding at a rate of 30 centimeters per year.
37. After four months, there are $300(4)^2 = 4800$ mussels in the bay. The population is growing at the rate $Z'(4)$ mussels per month, where $Z'(t) = 600t$, so the rate of increase is 2400 mussels per month.
41. The marginal cost of producing the 25th item is $C'(25)$, where $C'(q) = 4q$, so the marginal cost is \$100. This means that the cost of production increases by about \$100 when we add one unit to a production level of 25 units.

45. **(a)** The marginal cost function describes the change in cost on the margin, i.e. the change in cost associated with each additional item produced. Thus it is the change in cost with respect to change in production (dC/dq). We recognize this as an expression for the derivative of $C(q)$ with respect to q. Thus, the marginal cost function equals $C'(q) = 0.08(3q^2) + 75 = 0.24q^2 + 75$.

(b)

$$C(50) = 0.08(50)^3 + 75(50) + 1000 = \$14{,}750.$$

$C(50)$ tells us how much it costs to produce 50 items. From above we can see that the company spends \$14,750 to produce 50 items, and thus the units for $C(q)$ are dollars.

$$C'(50) = 0.24(50)^2 + 75 = \$675 \text{ per item}.$$

$C'(q)$ tells us the change in cost to produce one additional item of product. Thus at $q = 50$ costs will increase by \$675 for each additional item of product produced, and thus the units are dollars/item.

49. **(a)** Velocity $v(t) = \frac{dy}{dt} = \frac{d}{dt}(1250 - 16t^2) = -32t$.
Since $t \geq 0$, the ball's velocity is negative. This is reasonable, since its height y is decreasing.

(b) The ball hits the ground when its height $y = 0$. This gives

$$1250 - 16t^2 = 0$$
$$t \approx \pm 8.84 \text{ seconds}$$

We discard $t = -8.84$ because time t is nonnegative. So the ball hits the ground about 8.84 seconds after its release, at which time its velocity is

$$v(8.84) = -32(8.84) = -282.88 \text{ feet/sec} = -192.87 \text{ mph}.$$

Solutions for Section 3.2

1. $f'(x) = 2e^x + 2x$.

5. $\dfrac{dy}{dx} = 5 \cdot 5^t \ln 5 + 6 \cdot 6^t \ln 6$

9. $\dfrac{dy}{dx} = 5(\ln 2)(2^x) - 5$.

13. $y' = Ae^t$

17. $R' = \frac{3}{q}$.

21. $\dfrac{dy}{dx} = 2x + 4 - 3/x$.

25. $C(500) = 1000 + 300\ln(500) \approx 2864.38$; it costs about \$2864 to produce 500 units. $C'(q) = \frac{300}{q}$, $C'(500) = \frac{300}{500} = 0.6$. When the production level is 500, each additional unit costs about \$0.60 to produce.

29. $\dfrac{dV}{dt} = 75(1.35)^t \ln 1.35 \approx 22.5(1.35)^t$.

33. **(a)** For $y = \ln x$, we have $y' = 1/x$, so the slope of the tangent line is $f'(1) = 1/1 = 1$. The equation of the tangent line is $y - 0 = 1(x - 1)$, so, on the tangent line, $y = g(x) = x - 1$.

(b) Using a value on the tangent line to approximate $\ln(1,1)$, we have

$$\ln(1.1) \approx g(1.1) = 1.1 - 1 = 0.1.$$

Similarly, $\ln(2)$ is approximated by

$$\ln(2) \approx g(2) = 2 - 1 = 1.$$

(c) From Figure 3.1, we see that $f(1.1)$ and $f(2)$ are below $g(x) = x - 1$. Similarly, $f(0.9)$ and $f(0.5)$ are also below $g(x)$. This is true for any approximation of this function by a tangent line since f is concave down ($f''(x) = -\frac{1}{x^2} < 0$ for all $x > 0$). Thus, for a given x-value, the y-value given by the function is always below the value given by the tangent line.

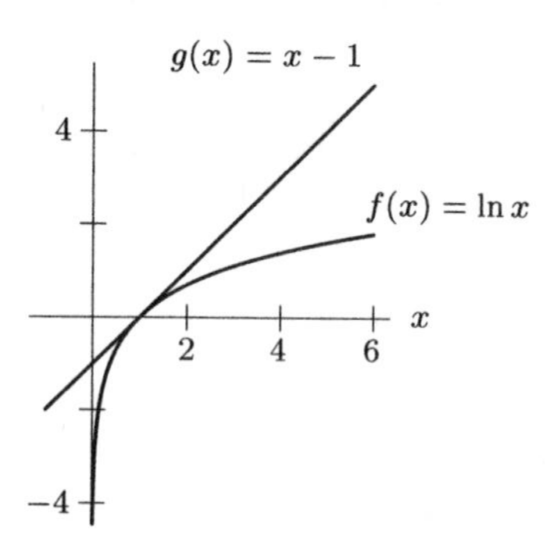

Figure 3.1

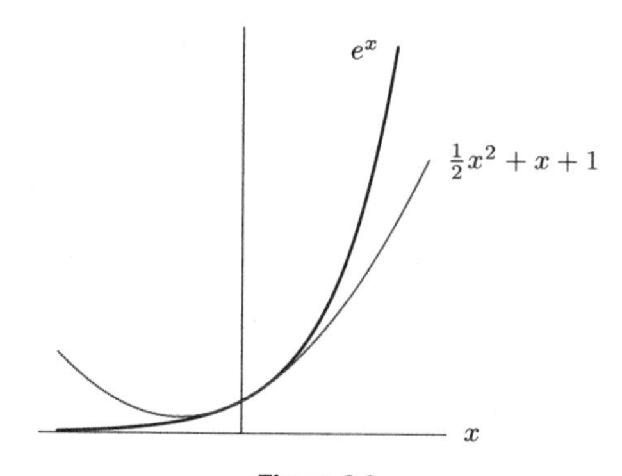

Figure 3.2

37.

$$g(x) = ax^2 + bx + c \qquad f(x) = e^x$$
$$g'(x) = 2ax + b \qquad f'(x) = e^x$$
$$g''(x) = 2a \qquad f''(x) = e^x$$

So, using $g''(0) = f''(0)$, etc., we have $2a = 1$, $b = 1$, and $c = 1$, and thus $g(x) = \frac{1}{2}x^2 + x + 1$, as shown in Figure 3.2.

The two functions do look very much alike near $x = 0$. They both increase for large values of x, but e^x increases much more quickly. For very negative values of x, the quadratic goes to ∞ whereas the exponential goes to 0. By choosing a function whose first few derivatives agreed with the exponential when $x = 0$, we got a function which looks like the exponential for x-values near 0.

Solutions for Section 3.3

1. $f'(x) = 99(x+1)^{98} \cdot 1 = 99(x+1)^{98}$.

5. $\dfrac{dw}{dr} = 3(5r-6)^2 \cdot 5 = 15(5r-6)^2$.

9. $y' = -4e^{-4t}$.

13. $\dfrac{dy}{dx} = -3(2x) + 2\left(3e^{3x}\right) = -6x + 6e^{3x}$.

17. $\dfrac{dw}{dt} = -6te^{-3t^2}$.

21. $f'(t) = \dfrac{2t}{t^2+1}$.

25. $g'(t) = \dfrac{1}{4t+9} \cdot 4 = \dfrac{4}{4t+9}$.

29. $\dfrac{dy}{dx} = 2(5+e^x)e^x$.

33.

$$f(x) = 6e^{5x} + e^{-x^2} \qquad f'(x) = 30e^{5x} - 2xe^{-x^2}$$
$$f(1) = 6e^5 + e^{-1} \qquad f'(1) = 30e^5 - 2(1)e^{-1}$$

$$\begin{aligned}
y - y_1 &= m(x - x_1)\\
y - (6e^5 + e^{-1}) &= (30e^5 - 2e^{-1})(x-1)\\
y - (6e^5 + e^{-1}) &= (30e^5 - 2e^{-1})x - (30e^5 - 2e^{-1})\\
y &= (30e^5 - 2e^{-1})x - 30e^5 + 2e^{-1} + 6e^5 + e^{-1}\\
&\approx 4451.66x - 3560.81.
\end{aligned}$$

37. The concentration of the drug in the body after 4 hours is

$$f(4) = 27e^{-0.14(4)} = 15.4 \text{ ng/ml}.$$

The rate of change of the concentration is the derivative

$$f'(t) = 27e^{-0.14t}(-0.14) = -3.78e^{-0.14t}.$$

At $t = 4$, the concentration is changing at a rate of

$$f'(4) = -3.78e^{-0.14(4)} = -2.16 \text{ ng/ml per hour}.$$

41. **(a)** $\dfrac{dB}{dt} = P\left(1 + \dfrac{r}{100}\right)^t \ln\left(1 + \dfrac{r}{100}\right)$. The expression $\dfrac{dB}{dt}$ tells us how fast the amount of money in the bank is changing with respect to time for fixed initial investment P and interest rate r.

(b) $\dfrac{dB}{dr} = Pt\left(1 + \dfrac{r}{100}\right)^{t-1} \dfrac{1}{100}$. The expression $\dfrac{dB}{dr}$ indicates how fast the amount of money changes with respect to the interest rate r, assuming fixed initial investment P and time t.

Solutions for Section 3.4

1. By the product rule, $f'(x) = 2x(x^3+5) + x^2(3x^2) = 2x^4 + 3x^4 + 10x = 5x^4 + 10x$. Alternatively, $f'(x) = (x^5 + 5x^2)' = 5x^4 + 10x$. The two answers should, and do, match.

5. $y' = 2^x + x(\ln 2)2^x = 2^x(1 + x\ln 2)$.

9. $\dfrac{dy}{dt} = 2te^t + (t^2+3)e^t = e^t(t^2 + 2t + 3)$.

13. Since $f(t) = 5t^{-1} + 6t^{-2}$, we have

$$f'(t) = -5t^{-2} - 12t^{-3} = -\frac{5}{t^2} - \frac{12}{t^3}.$$

17. $f'(z) = \dfrac{1}{2\sqrt{z}}e^{-z} - \sqrt{z}e^{-z}$.

21. $w' = (3t^2+5)(t^2 - 7t + 2) + (t^3 + 5t)(2t - 7)$.

25. Using the quotient rule,

$$\frac{dy}{dx} = \frac{d}{dx}\left(\frac{e^x}{1+e^x}\right) = \frac{e^x(1+e^x) - e^x(e^x)}{(1+e^x)^2} = \frac{e^x + e^{2x} - e^{2x}}{(1+e^x)^2} = \frac{e^x}{(1+e^x)^2}.$$

29. Since a and b are constants, we have $f'(x) = 3(ax^2+b)^2(2ax) = 6ax(ax^2+b)^2$.

33.

$$f'(x) = 3(2x-5) + 2(3x+8) = 12x + 1$$
$$f''(x) = 12.$$

37. To find the equation of the line tangent to the graph of $P(t) = t\ln t$ at $t = 2$ we must find the point $(2, P(2))$ as well as the slope of the tangent line at $t = 2$. $P(2) = 2(\ln 2) \approx 1.386$. Thus we have the point $(2, 1.386)$. To find the slope, we must first find $P'(t)$:

$$P'(t) = t\frac{1}{t} + \ln t(1) = 1 + \ln t.$$

At $t = 2$ we have

$$P'(2) = 1 + \ln 2 \approx 1.693$$

Since we now have the slope of the line and a point, we can solve for the equation of the line:

$$Q(t) - 1.386 = 1.693(t-2)$$
$$Q(t) - 1.386 = 1.693t - 3.386$$
$$Q(t) = 1.693t - 2.$$

The equation of the tangent line is $Q(t) = 1.693t - 2$. We see our results displayed graphically in Figure 3.3.

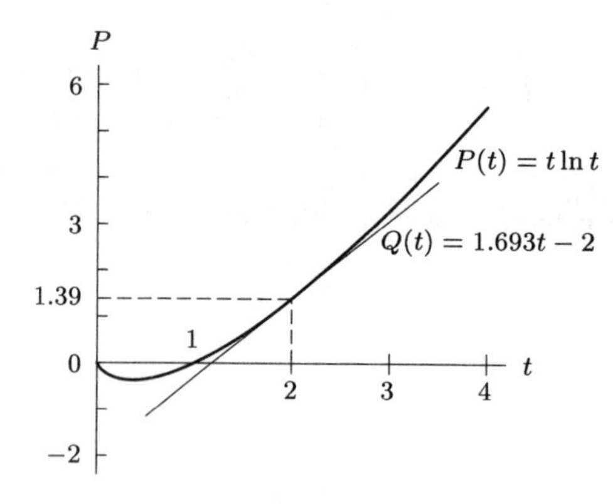

Figure 3.3

Solutions for Section 3.5

1. $\frac{dy}{dx} = 5\cos x$.

5. $\frac{dy}{dx} = 5\cos x - 5$.

9. $\frac{dy}{dt} = A(\cos(Bt)) \cdot B = AB\cos(Bt)$.

13. $\frac{dy}{dt} = 6 \cdot 2\cos(2t) + (-4\sin(4t)) = 12\cos(2t) - 4\sin(4t)$.

17. Using the quotient rule, we get

$$\frac{d}{dt}\left(\frac{t^2}{\cos t}\right) = \frac{2t\cos t - t^2(-\sin t)}{(\cos t)^2} = \frac{2t\cos t + t^2\sin t}{(\cos t)^2}.$$

21. We begin by taking the derivative of $y = \sin(x^4)$ and evaluating at $x = 10$:

$$\frac{dy}{dx} = \cos(x^4) \cdot 4x^3.$$

Evaluating $\cos(10{,}000)$ on a calculator (in radians) we see $\cos(10{,}000) < 0$, so we know that $dy/dx < 0$, and therefore the function is decreasing.
Next, we take the second derivative and evaluate it at $x = 10$, giving $\sin(10{,}000) < 0$:

$$\frac{d^2y}{dx^2} = \underbrace{\cos(x^4) \cdot (12x^2)}_{\text{negative}} + \underbrace{4x^3 \cdot (-\sin(x^4))(4x^3)}_{\text{positive, but much larger in magnitude}}.$$

From this we can see that $d^2y/dx^2 > 0$, thus the graph is concave up.

25. Since P is a function of r and r is a function of t, we use the chain rule. Taking the derivative of P with respect to r we have

$$\frac{dP}{dr} = \frac{(k+r)\cdot c - cr \cdot 1}{(k+r)^2} = \frac{kc}{(k+r)^2}.$$

Taking the derivative of r with respect to t we have

$$\frac{dr}{dt} = \frac{\pi}{6} \cdot 10\cos\left(\frac{\pi}{6}t\right) = \frac{5\pi}{3}\cos\left(\frac{\pi}{6}t\right).$$

Thus, the derivative of P with respect to t is

$$\frac{dP}{dt} = \frac{dP}{dr} \cdot \frac{dr}{dt} = \frac{5\pi}{3}\frac{kc}{(k + 10\sin\left(\frac{\pi}{6}t\right) + 10)^2}\cos\left(\frac{\pi}{6}t\right).$$

Solutions for Chapter 3 Review

1. $f'(t) = 24t^3$.

5. $f'(x) = 3x^2 - 6x + 5$.

9. $f'(x) = 2x + \frac{3}{x}$.

13. $f'(z) = \frac{1}{z^2+1}(2z) = \frac{2z}{z^2+1}$.

17. $s'(t) = 2t + \frac{2}{t}$.

21. $f'(x) = 2\cos(2x)$.

25. We use the chain rule with $z = g(x) = 1 + e^x$ as the inside function and $f(z) = \ln z$ as the outside function. Since $g'(x) = e^x$ and $f'(z) = 1/z$ we have

$$\frac{d}{dx}(\ln(1+e^x)) = \frac{1}{z}e^x = \frac{e^x}{1+e^x}.$$

29. This is a quotient where $u(x) = 1 + e^x$ and $v(x) = 1 - e^{-x}$ so that $q(x) = u(x)/v(x)$.
Using the quotient rule the derivative is

$$q'(x) = \frac{vu' - uv'}{v^2},$$

where $u' = e^x$ and $v' = e^{-x}$. Therefore

$$q'(x) = \frac{(1-e^{-x})e^x - e^{-x}(1+e^x)}{(1-e^{-x})^2} = \frac{e^x - 2 - e^{-x}}{(1-e^{-x})^2}.$$

33. We use the chain rule with $z = g(x) = x^3$ as the inside function and $f(z) = \cos z$ as the outside function. Since $g'(x) = 3x^2$ and $f'(z) = -\sin z$, we have

$$\frac{d}{dx}\left(\cos(x^3)\right) = -\sin z \cdot (3x^2) = -3x^2\sin(x^3).$$

37. Since $f(1) = 2(1^3) - 5(1^2) + 3(1) - 5 = -5$, the point $(1, -5)$ is on the line. We use the derivative to find the slope. Differentiating gives

$$f'(x) = 6x^2 - 10x + 3,$$

and so the slope at $x = 1$ is

$$f'(1) = 6(1^2) - 10(1) + 3 = -1.$$

The equation of the tangent line is

$$\begin{aligned} y - (-5) &= -1(x-1) \\ y + 5 &= -x + 1 \\ y &= -4 - x. \end{aligned}$$

41. $R(p) = p \cdot q = 5000pe^{-0.08p}$

$R(10) = 50{,}000e^{-0.8} \approx 22{,}466$; revenues of about \$22,466 can be expected when the selling price is \$10.
$R'(p) = 5000e^{-0.08p} + 5000p(-0.08)e^{-0.08p} = (5000 - 400p)e^{-0.08p}$

$R'(10) = 1000e^{-0.8} \approx 449$; if price is increased by one dollar over \$10, revenue will increase by about \$449.

45. **(a)** $H'(2) = r'(2) + s'(2) = -1 + 3 = 2.$
(b) $H'(2) = 5s'(2) = 5(3) = 15.$
(c) $H'(2) = r'(2)s(2) + r(2)s'(2) = -1 \cdot 1 + 4 \cdot 3 = 11.$
(d) $H'(2) = \dfrac{r'(2)}{2\sqrt{r(2)}} = \dfrac{-1}{2\sqrt{4}} = -\dfrac{1}{4}.$

49. Decreasing means $f'(x) < 0$:

$$f'(x) = 4x^3 - 12x^2 = 4x^2(x-3),$$

so $f'(x) < 0$ when $x < 3$ and $x \neq 0$. Concave up means $f''(x) > 0$:

$$f''(x) = 12x^2 - 24x = 12x(x-2)$$

so $f''(x) > 0$ when

$$\begin{gathered} 12x(x-2) > 0 \\ x < 0 \quad \text{or} \quad x > 2. \end{gathered}$$

So, both conditions hold for $x < 0$ or $2 < x < 3$.

53. Since W is proportional to r^3, we have $W = kr^3$ for some constant k. Thus, $dW/dr = k(3r^2) = 3kr^2$. Thus, dW/dr is proportional to r^2.

57. If $f(x) = x^n$, then $f'(x) = nx^{n-1}$. This means $f'(1) = n \cdot 1^{n-1} = n \cdot 1 = n$, because any power of 1 equals 1.

Solutions to Problems on Establishing the Derivative Formulas

1. Using the definition of the derivative, we have

$$\begin{aligned} f'(x) &= \lim_{h\to 0} \frac{f(x+h)-f(x)}{h} \\ &= \lim_{h\to 0} \frac{2(x+h)+1-(2x+1)}{h} \\ &= \lim_{h\to 0} \frac{2x+2h+1-2x-1}{h} \\ &= \lim_{h\to 0} \frac{2h}{h}. \end{aligned}$$

As long as h is very close to, but not actually equal to, zero we can say that $\lim_{h\to 0} \frac{2h}{h} = 2$, and thus conclude that $f'(x) = 2$.

5. The definition of the derivative states that

$$f'(x) = \lim_{h\to 0} \frac{f(x+h)-f(x)}{h}.$$

Using this definition, we have

$$\begin{aligned} f'(x) &= \lim_{h\to 0} \frac{4(x+h)^2+1-(4x^2+1)}{h} \\ &= \lim_{h\to 0} \frac{4x^2+8xh+4h^2+1-4x^2-1}{h} \\ &= \lim_{h\to 0} \frac{8xh+4h^2}{h} \\ &= \lim_{h\to 0} \frac{h(8x+4h)}{h}. \end{aligned}$$

As long as h approaches, but does not equal, zero we can cancel h in the numerator and denominator. The derivative now becomes

$$\lim_{h\to 0}(8x+4h) = 8x.$$

Thus, $f'(x) = 6x$ as we stated above.

9. Since $f(x) = C$ for all x, we have $f(x+h) = C$. Using the definition of the derivative, we have

$$\begin{aligned} f'(x) &= \lim_{h\to 0} \frac{f(x+h)-f(x)}{h} \\ &= \lim_{h\to 0} \frac{C-C}{h} \\ &= \lim_{h\to 0} \frac{0}{h}. \end{aligned}$$

As h gets very close to zero without actually equaling zero, we have $0/h = 0$, so

$$f'(x) = \lim_{h\to 0}(0) = 0.$$

Solutions to Practice Problems on Differentiation

1. $f'(t) = 2t + 4t^3$

5. $f'(x) = -2x^{-1} + 5\left(\frac{1}{2}x^{-1/2}\right) = \dfrac{-2}{x} + \dfrac{5}{2\sqrt{x}}$

9. $D'(p) = 2pe^{p^2} + 10p$

13. $s'(t) = \dfrac{16}{2t+1}$

17. $C'(q) = 3(2q+1)^2 \cdot 2 = 6(2q+1)^2$

21. $y' = 2x\ln(2x+1) + \dfrac{2x^2}{2x+1}$

25. $g'(t) = 15\cos(5t)$

29. $y' = 17 + 12x^{-1/2}$.

33. Either notice that $f(x) = \dfrac{x^2+3x+2}{x+1}$ can be written as $f(x) = \dfrac{(x+2)(x+1)}{x+1}$ which reduces to $f(x) = x+2$, giving $f'(x) = 1$, or use the quotient rule which gives

$$\begin{aligned} f'(x) &= \frac{(x+1)(2x+3)-(x^2+3x+2)}{(x+1)^2} \\ &= \frac{2x^2+5x+3-x^2-3x-2}{(x+1)^2} \\ &= \frac{x^2+2x+1}{(x+1)^2} \\ &= \frac{(x+1)^2}{(x+1)^2} \\ &= 1. \end{aligned}$$

37. $q'(r) = \dfrac{3(5r+2)-3r(5)}{(5r+2)^2} = \dfrac{15r+6-15r}{(5r+2)^2} = \dfrac{6}{(5r+2)^2}$

41. $h'(w) = 5(w^4-2w)^4(4w^3-2)$

45. $h'(w) = 6w^{-4} + \dfrac{3}{2}w^{-1/2}$

49. Using the chain rule, $g'(\theta) = (\cos\theta)e^{\sin\theta}$.

53. $h'(r) = \dfrac{d}{dr}\left(\dfrac{r^2}{2r+1}\right) = \dfrac{(2r)(2r+1)-2r^2}{(2r+1)^2} = \dfrac{2r(r+1)}{(2r+1)^2}$.

57. $f'(x) = \dfrac{3x^2}{9}(3\ln x - 1) + \dfrac{x^3}{9}\left(\dfrac{3}{x}\right) = x^2\ln x - \dfrac{x^2}{3} + \dfrac{x^2}{3} = x^2\ln x$

61. Using the quotient rule gives

$$\begin{aligned} w'(r) &= \frac{2ar(b+r^3)-3r^2(ar^2)}{(b+r^3)^2} \\ &= \frac{2abr-ar^4}{(b+r^3)^2}. \end{aligned}$$

CHAPTER FOUR

Solutions for Section 4.1

1. We find a critical point by noting where $f'(x) = 0$ or f' is undefined. Since the curve is smooth throughout, f' is always defined, so we look for where $f'(x) = 0$, or equivalently where the tangent line to the graph is horizontal. These points are shown in Figure 4.1:

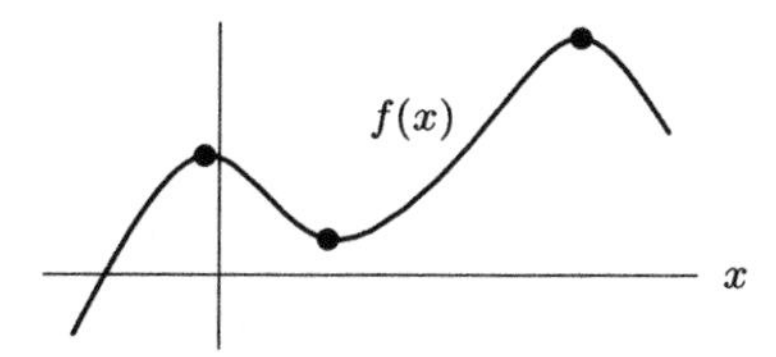

Figure 4.1

As we can see, there are three critical points. The leftmost one is a local maximum, because points near it are all lower; similarly, the middle critical point is surrounded by higher points, and is a local minimum. The critical point to the right is a local maximum.

5. Since $f'(x) = 4x^3 - 12x^2 + 8$, we see that $f'(1) = 0$, as we expected. We apply the second derivative test to $f''(x) = 12x^2 - 24x$. Since $f''(1) = -12 < 0$, the graph is concave down at the critical point $x = 1$, making it a local maximum.

9.

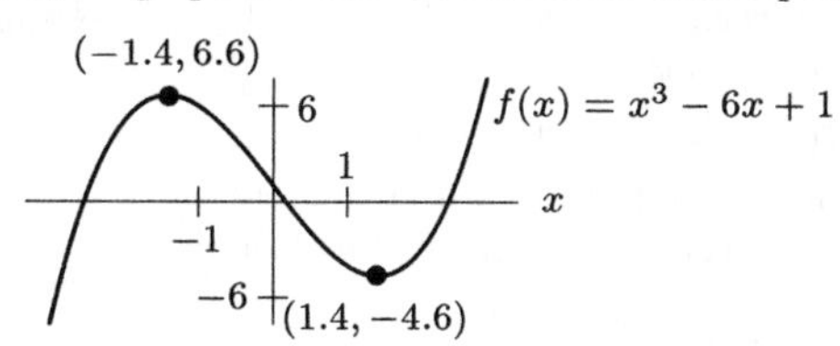

The graph of f above appears to be increasing for $x < -1.4$, decreasing for $-1.4 < x < 1.4$, and increasing for $x > 1.4$. There is a local maximum near $x = -1.4$ and local minimum near $x = 1.4$. The derivative of f is $f'(x) = 3x^2 - 6$. Thus $f'(x) = 0$ when $x^2 = 2$, that is $x = \pm\sqrt{2}$. This explains the critical points near $x = \pm 1.4$. Since $f'(x)$ changes from positive to negative at $x = -\sqrt{2}$, and from negative to positive at $x = \sqrt{2}$, there is a local maximum at $x = -\sqrt{2}$ and a local minimum at $x = \sqrt{2}$.

13. The graph of f above appears to be decreasing for $0 < x < 0.37$, and then increasing for $x > 0.37$. We have $f'(x) = \ln x + x(1/x) = \ln x + 1$, so $f'(x) = 0$ when $\ln x = -1$, that is, $x = e^{-1} \approx 0.37$. This is the only place where f' changes sign and $f'(1) = 1 > 0$, so the graph must decrease for $0 < x < e^{-1}$ and increase for $x > e^{-1}$. Thus, there is a local minimum at $x = e^{-1}$.

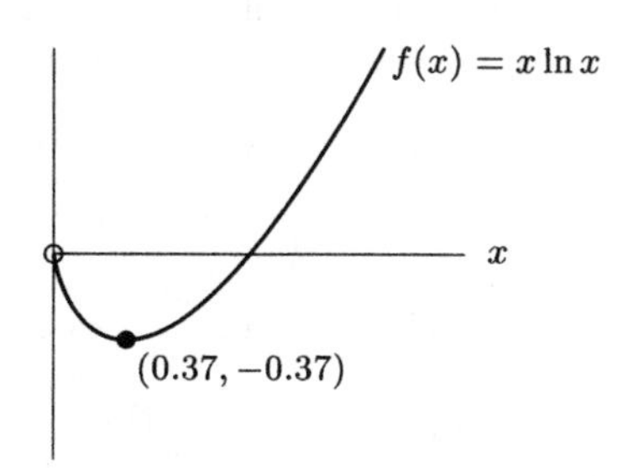

17. **(a)** One possible answer is shown in Figure 4.2.
(b) One possible answer is shown in Figure 4.3

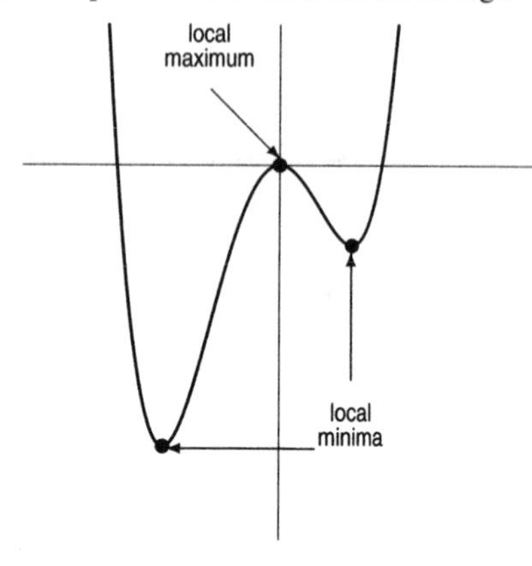

Figure 4.2

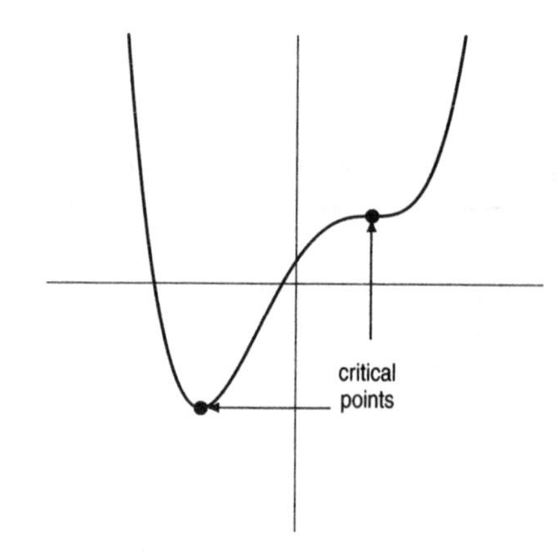

Figure 4.3

21. Using the product rule on the function $f(x) = axe^{bx}$, we have $f'(x) = ae^{bx} + abxe^{bx} = ae^{bx}(1 + bx)$. We want $f(\frac{1}{3}) = 1$, and since this is to be a maximum, we require $f'(\frac{1}{3}) = 0$. These conditions give

$$f(1/3) = a(1/3)e^{b/3} = 1,$$
$$f'(1/3) = ae^{b/3}(1 + b/3) = 0.$$

Since $ae^{(1/3)b}$ is non-zero, we can divide both sides of the second equation by $ae^{(1/3)b}$ to obtain $0 = 1 + \frac{b}{3}$. This implies $b = -3$. Plugging $b = -3$ into the first equation gives us $a(\frac{1}{3})e^{-1} = 1$, or $a = 3e$. How do we know we have a maximum at $x = \frac{1}{3}$ and not a minimum? Since $f'(x) = ae^{bx}(1 + bx) = (3e)e^{-3x}(1 - 3x)$, and $(3e)e^{-3x}$ is always positive, it follows that $f'(x) > 0$ when $x < \frac{1}{3}$ and $f'(x) < 0$ when $x > \frac{1}{3}$. Since f' is positive to the left of $x = \frac{1}{3}$ and negative to the right of $x = \frac{1}{3}$, $f(\frac{1}{3})$ is a local maximum.

25. **(a)** In Figure 4.4, we see that $f(\theta) = \theta - \sin\theta$ has a zero at $\theta = 0$. To see if it has any other zeros near the origin, we use our calculator to zoom in. (See Figure 4.5.) No extra root seems to appear no matter how close to the origin we zoom. However, zooming can never tell you for sure that there is not a root that you have not found yet.

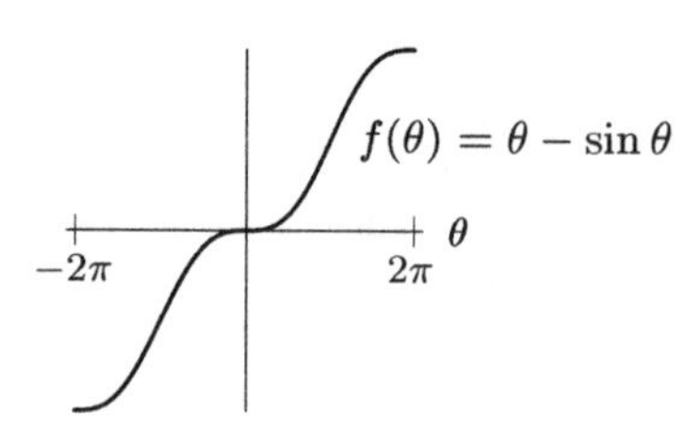

Figure 4.4: Graph of $f(\theta)$

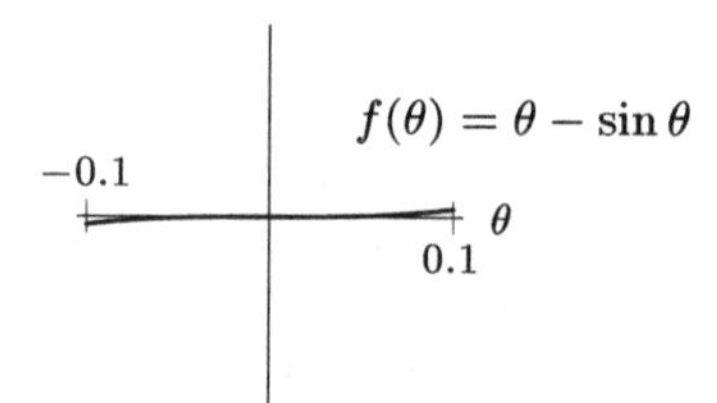

Figure 4.5: Graph of $f(\theta)$ Zoomed In

(b) Using the derivative, $f'(\theta) = 1 - \cos\theta$, we can argue that there is no other zero. Since $\cos\theta < 1$ for $0 < \theta \le 1$, we know $f'(\theta) > 0$ for $0 < \theta \le 1$. Thus, f increases for $0 < \theta \le 1$. Consequently, we conclude that the only zero of f is the one at the origin. If f had another zero at x_0, with $x_0 > 0$, then f would have to "turn around", and recross the x-axis at x_0. But if this were the case, f' would be nonpositive somewhere, which we know is not the case.

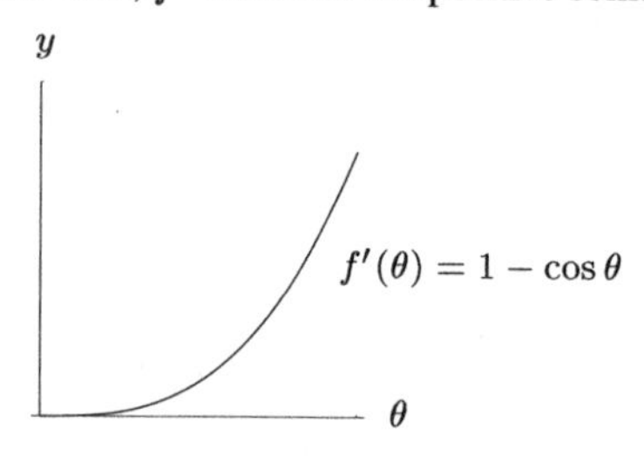

Figure 4.6: Graph of $f'(\theta)$

Solutions for Section 4.2

1. We find an inflection point by noting where the concavity changes. Such points are shown in Figure 4.7: There are two inflection points.

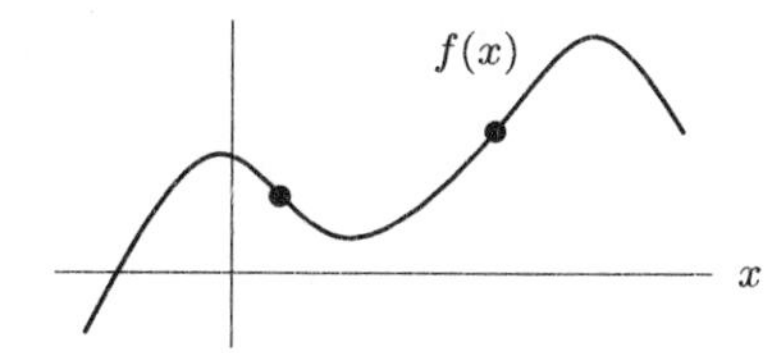

Figure 4.7

5. From the graph of $f(x)$ in the figure below, we see that the function must have two inflection points. We calculate $f'(x) = 4x^3 + 3x^2 - 6x$, and $f''(x) = 12x^2 + 6x - 6$. Solving $f''(x) = 0$ we find that:

$$x_1 = -1 \quad \text{and} \quad x_2 = \frac{1}{2}.$$

Since $f''(x) > 0$ for $x < x_1$, $f''(x) < 0$ for $x_1 < x < x_2$, and $f''(x) > 0$ for $x_2 < x$, it follows that both points are inflection points.

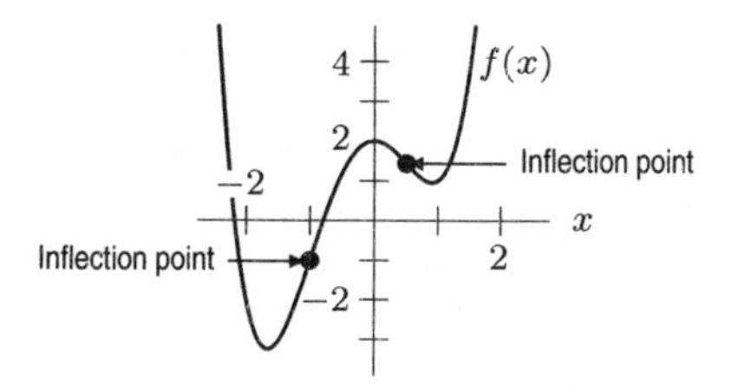

9. The derivative of $f(x)$ is $f'(x) = 4x^3 - 16x$. The critical points of $f(x)$ are points at which $f'(x) = 0$. Factoring, we get

$$4x^3 - 16x = 0$$
$$4x(x^2 - 4) = 0$$
$$4x(x-2)(x+2) = 0$$

So the critical points of f will be $x = 0, x = 2,$ and $x = -2$.

To find the inflection points of f we look for the points at which $f''(x)$ changes sign. At any such point $f''(x)$ is either zero or undefined. Since $f''(x) = 12x^2 - 16$ our candidate points are $x = \pm 2/\sqrt{3}$. At both of these points $f''(x)$ changes sign, so both of these points are inflection points.

From Figure 4.8, we see that the critical points $x = -2$ and $x = 2$ are local minima and the critical point $x = 0$ is a local maximum.

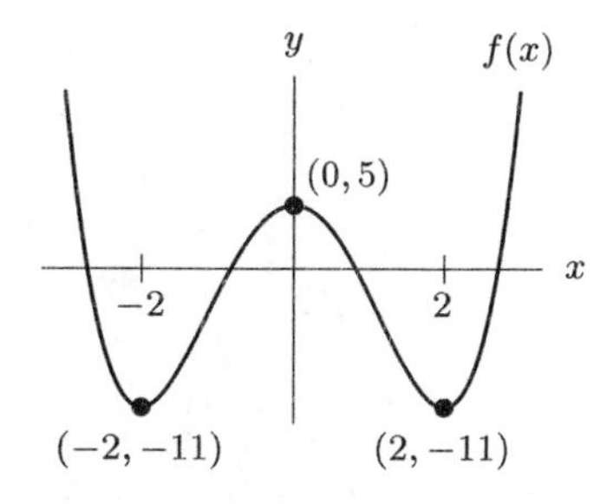

Figure 4.8

13. **(a)** One possible answer is shown in Figure 4.9.

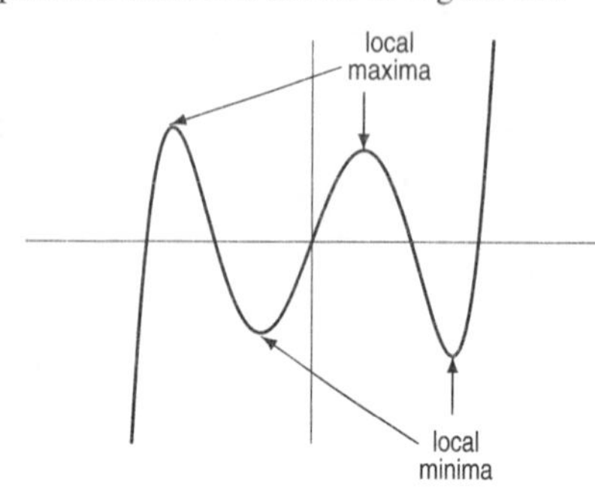

Figure 4.9

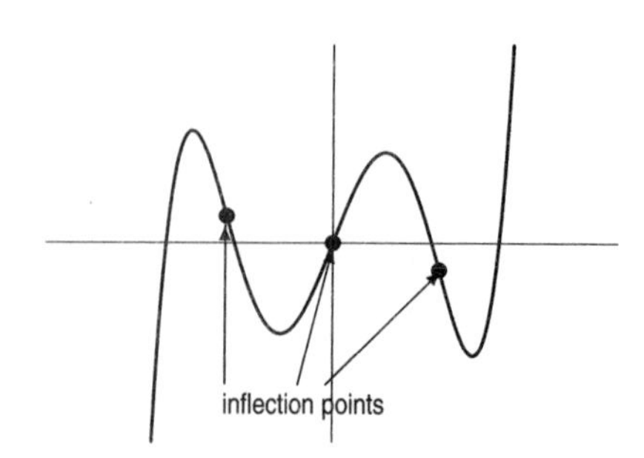

Figure 4.10

(b) This function is concave down at each local maximum and concave up at each local minimum, so it changes concavity at least three times. This function has at least 3 inflection points. See Figure 4.10

17. We have $f'(x) = 3x^2 - 36x - 10$ and $f''(x) = 6x - 36$. The inflection point occurs where $f''(x) = 0$, hence $6x - 36 = 0$. The inflection point is at $x = 6$. A graph is shown in Figure 4.11.

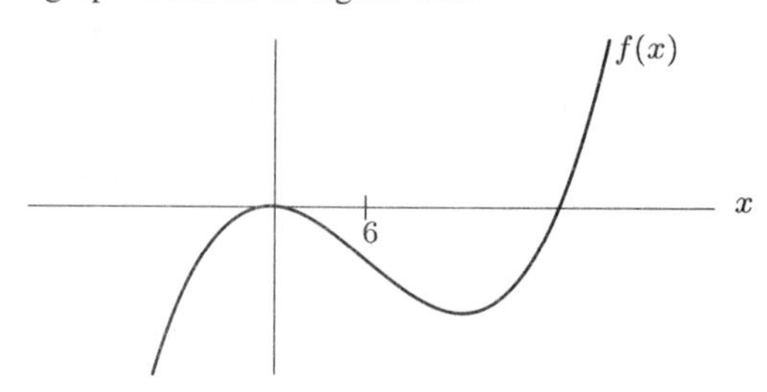

Figure 4.11

21.

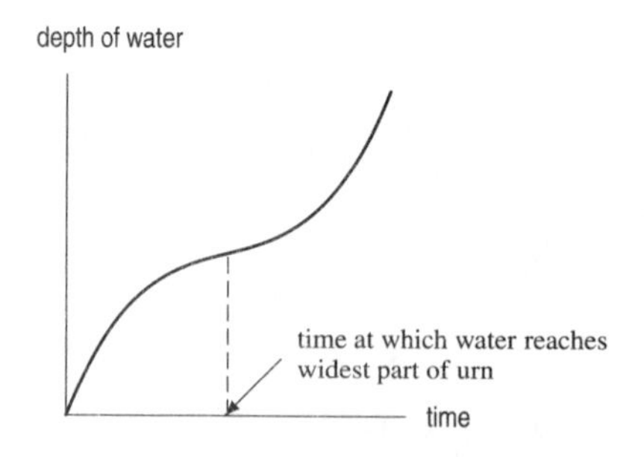

25. The local maxima and minima of f correspond to places where f' is zero and changes sign or, possibly, to the endpoints of intervals in the domain of f. The points at which f changes concavity correspond to local maxima and minima of f'. The change of sign of f', from positive to negative corresponds to a maximum of f and change of sign of f' from negative to positive corresponds to a minimum of f.

29.

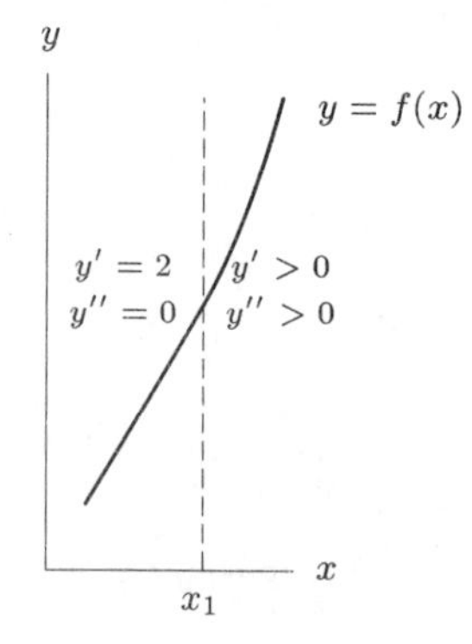

Solutions for Section 4.3

1.

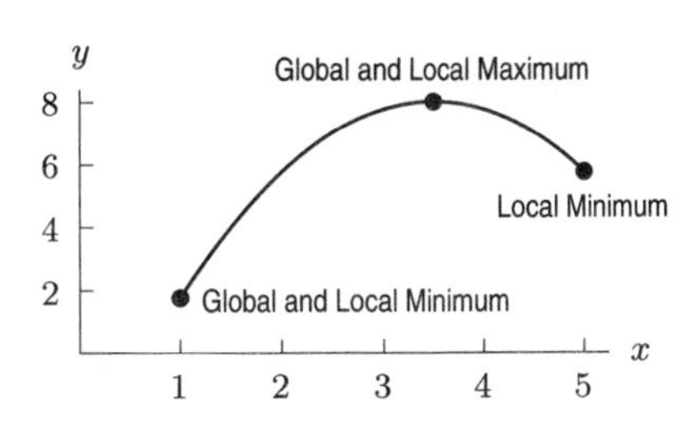

5. See Figure 4.12.

9. See Figure 4.13.

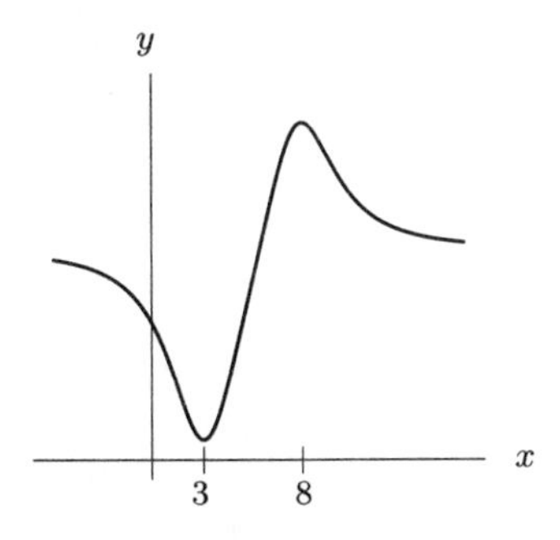

Figure 4.12

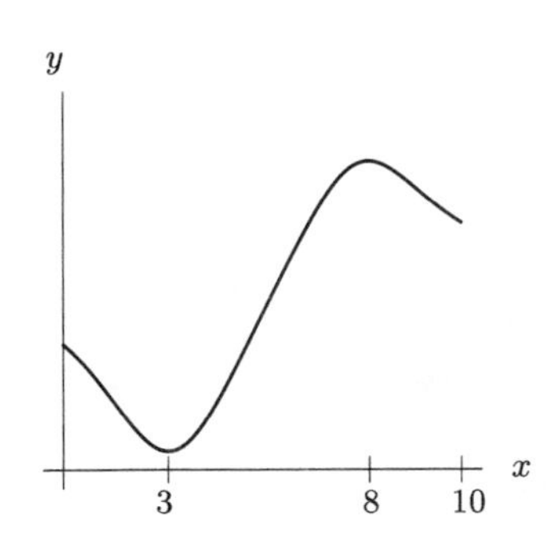

Figure 4.13

13. **(a)** $f'(x) = 1 - 1/x$. This is zero only when $x = 1$. Now $f'(x)$ is positive when $1 < x \le 2$, and negative when $0.1 < x < 1$. Thus $f(1) = 1$ is a local minimum. The endpoints $f(0.1) \approx 2.4026$ and $f(2) \approx 1.3069$ are local maxima.

(b) Comparing values of f shows that $x = 0.1$ gives the global maximum and $x = 1$ gives the global minimum.

17. To find the minimum value of E, we solve

$$\frac{dE}{dF} = 0.25 - \frac{2(1.7)}{F^3} = 0$$

$$F^3 = \frac{2(1.7)}{0.25}$$

$$F = \left(\frac{2(1.7)}{0.25}\right)^{1/3} = 2.4 \text{ hours.}$$

Since dE/dF is negative for $F < 2.4$ and dE/dF is positive for $F > 2.4$, a foraging time of $F = 2.4$ hours gives a local minimum. This is the global minimum for $F > 0$. See Figure 4.14.

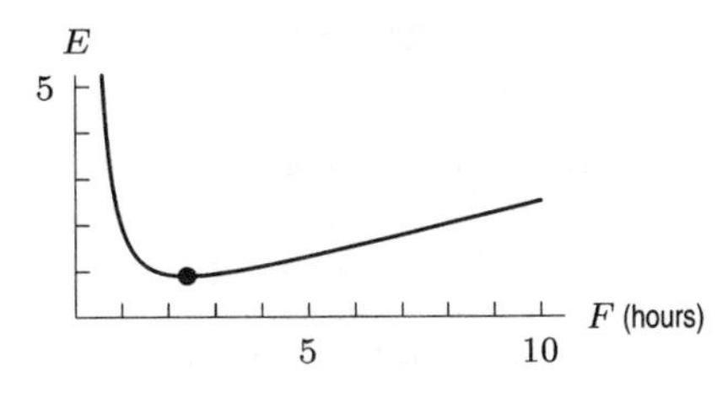

Figure 4.14

21. **(a)** When $t = 0$, we have $q(0) = 20(e^{-0} - e^{-2\cdot 0}) = 20(1-1) = 0$. This is as we would expect: initially there is none of the drug in the bloodstream.

(b) The maximum value of $q(t)$ occurs where

$$\begin{aligned} q'(t) = 20(-e^{-t} + 2e^{-2t}) &= 0 \\ -e^{-t} &= -2e^{-2t} \\ \frac{e^{-t}}{e^{-2t}} &= 2 \\ e^{t} &= 2 \\ t &= \ln 2 = 0.69 \text{ hours.} \end{aligned}$$

When $t = 0.69$ hours, we have $q(0.69) = 5$ mg.

(c) To see what happens in the long run, we let $t \to \infty$. Since $e^{-t} \to 0$ and $e^{-2t} \to 0$, we see that $q(t) \to 0$ as $t \to \infty$. This is as we would expect: in the long run, the drug will be metabolized or excreted and leave the body.

25. **(a)** The variables are S and H. At the maximum value of S, using the product rule, we have

$$\begin{aligned} \frac{dS}{dH} = ae^{-bH} + aH(-b)e^{-bH} &= 0 \\ ae^{-bH}(1 - bH) &= 0 \\ bH &= 1 \\ H &= \frac{1}{b}. \end{aligned}$$

Thus, the maximum value of S occurs when $H = 1/b$. To find the maximum value of S, we substitute $H = 1/b$, giving

$$S = a\frac{1}{b}e^{-b(1/b)} = \frac{a}{b}e^{-1}.$$

(b) Increasing a increases the maximum value of S. Increasing b decreases the maximum value of S.

29. **(a)** Suppose the farmer plants x trees per km^2. Then, for $x \leq 200$, the yield per tree is 400 kg, so

$$\text{Total yield, } y = 400x \text{ kg.}$$

For $x > 200$, the yield per tree is reduced by 1 kg for each tree over 200, so the yield per tree is $400 - (x - 200)$ kg $= (600 - x)$ kg. Thus,

$$\text{Total yield, } y = (600 - x)x = 600x - x^2 \text{ kg.}$$

The graph of total yield is in Figure 4.15. It is a straight line for $x \leq 200$ and a parabola for $x > 200$.

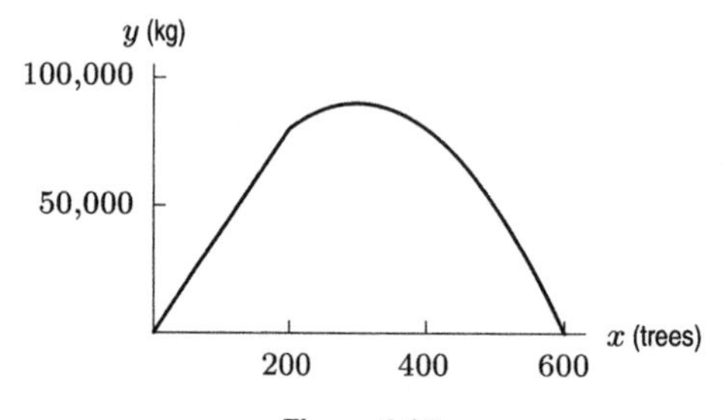

Figure 4.15

(b) The maximum occurs where the derivative of the quadratic is zero, or

$$\begin{aligned} \frac{dy}{dx} = 600 - 2x &= 0 \\ x &= 300 \text{ trees/km}^2. \end{aligned}$$

33. **(a)** Differentiating using the chain rule gives

$$p'(x) = \frac{1}{\sigma\sqrt{2\pi}}e^{-(x-\mu)^2/(2\sigma^2)} \cdot \left(-2\frac{(x-\mu)}{2\sigma^2}\right) = -\frac{(x-\mu)}{\sigma^3\sqrt{2\pi}}e^{-(x-\mu)^2/(2\sigma^2)}.$$

So $p'(x) = 0$ where $x - \mu = 0$, so $x = \mu$.

Differentiating again using the product rule gives

$$p''(x) = \frac{-1}{\sigma^3\sqrt{2\pi}}\left(1 \cdot e^{-(x-\mu)^2/(2\sigma^2)} + (x-\mu)e^{-(x-\mu)^2/(2\sigma^2)} \cdot \left(\frac{-2(x-\mu)}{2\sigma^2}\right)\right)$$
$$= -\frac{e^{-(x-\mu)^2/(2\sigma^2)}}{\sigma^5\sqrt{2\pi}}(\sigma^2 - (x-\mu)^2).$$

Substituting $x = \mu$ gives

$$p''(\mu) = -\frac{e^0}{\sigma^5\sqrt{2\pi}}\sigma^2 = -\frac{1}{\sigma^3\sqrt{2\pi}}.$$

Since $p''(\mu) < 0$, there is a local maximum at $x = \mu$. Since this local maximum is the only critical point, it is a global maximum. See Figure 4.16.

(b) Using the formula for $p''(x)$, we see that $p''(x) = 0$ where

$$\sigma^2 - (x-\mu)^2 = 0$$
$$x - \mu = \pm\sigma$$
$$x = \mu \pm \sigma.$$

Since $p''(x)$ changes sign at $x = \mu + \sigma$ and $x = \mu - \sigma$, these are both points of inflection. See Figure 4.16.

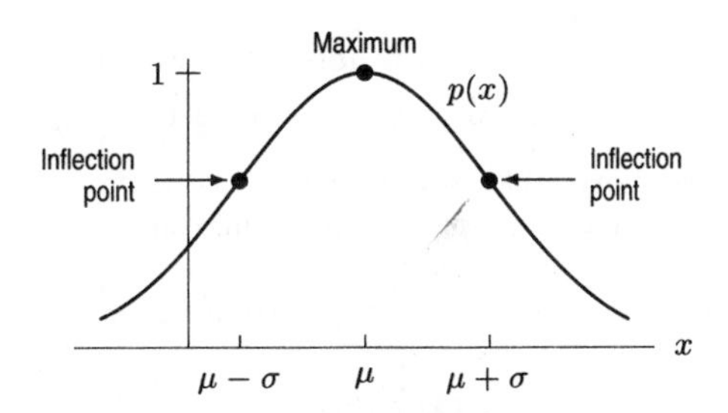

Figure 4.16

Solutions for Section 4.4

1. **(a)** Profit is maximized when $R(q) - C(q)$ is as large as possible. This occurs at $q = 2500$, where profit $= 7500 - 5500 =$ \$2000.

(b) We see that $R(q) = 3q$ and so the price is $p = 3$, or \$3 per unit.

(c) Since $C(0) = 3000$, the fixed costs are \$3000.

5. The profit function is positive when $R(q) > C(q)$, and negative when $C(q) > R(q)$. It's positive for $5.5 < q < 12.5$, and negative for $0 < q < 5.5$ and $12.5 < q$. Profit is maximized when $R(q) > C(q)$ and $R'(q) = C'(q)$ which occurs at about $q = 9.5$. See Figure 4.17.

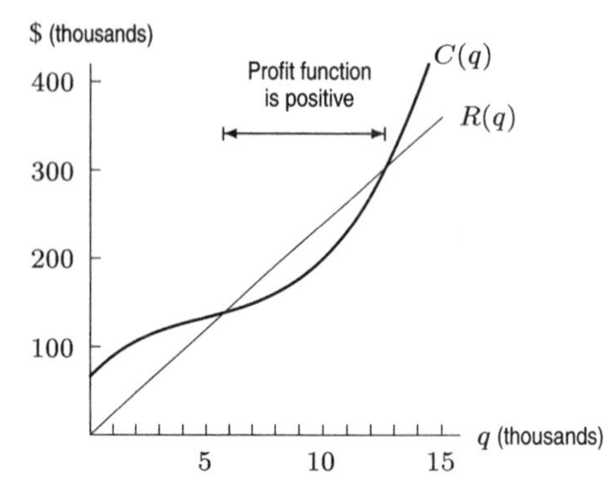

Figure 4.17

9. We first need to find an expression for $R(q)$, or revenue in terms of quantity sold. We know that $R(q) = pq$, where p is the price of one item. Here $p = 45 - 0.01q$, so we make the substitution

$$R(q) = (45 - .01q)q = 45q - 0.01q^2.$$

This is the function we want to maximize. Finding the derivative and setting it equal to 0 yields

$$\begin{aligned} R'(q) &= 0 \\ 45 - 0.02q &= 0 \\ 0.02q &= 45 \text{ so} \\ q &= 2250. \end{aligned}$$

Is this a maximum?

$$\begin{aligned} R'(q) &> 0 \text{ for } q < 2250 \text{ and} \\ R'(q) &< 0 \text{ for } q > 2250. \end{aligned}$$

So we conclude that $R(q)$ has a local maximum at $q = 2250$. Testing $q = 0$, the only endpoint, $R(0) = 0$, which is less than $R(2250) = \$50{,}625$. So we conclude that revenue is maximized at $q = 2250$. The price of each item at this production level is

$$p = 45 - .01(2250) = \$22.50$$

and total revenue is

$$pq = \$22.50(2250) = \$50{,}625,$$

which agrees with the above answer.

13. We first need to find an expression for revenue in terms of price. At a price of \$8, 1500 tickets are sold. For each \$1 above \$8, 75 fewer tickets are sold. This suggests the following formula for q, the quantity sold for any price p.

$$\begin{aligned} q &= 1500 - 75(p - 8) \\ &= 1500 - 75p + 600 \\ &= 2100 - 75p. \end{aligned}$$

We know that $R = pq$, so substitution yields

$$R(p) = p(2100 - 75p) = 2100p - 75p^2$$

To maximize revenue, we find the derivative of $R(p)$ and set it equal to 0.

$$\begin{aligned} R'(p) &= 2100 - 150p = 0 \\ 150p &= 2100 \end{aligned}$$

so $p = \frac{2100}{150} = 14$. Does $R(p)$ have a maximum at $p = 14$? Using the first derivative test,

$$\begin{aligned} R'(p) &> 0 \text{ if } p < 14 \text{ and} \\ R'(p) &< 0 \text{ if } p > 14. \end{aligned}$$

So $R(p)$ has a local maximum at $p = 14$. Since this is the only critical point for $p \geq 0$, it must be a global maximum. So we conclude that revenue is maximized when the price is \$14.

17. (a) We know that Profit = Revenue − Cost, so differentiating with respect to q gives:

$$\text{Marginal Profit} = \text{Marginal Revenue} - \text{Marginal Cost}.$$

We see from the figure in the problem that just to the left of $q = a$, marginal revenue is less than marginal cost, so marginal profit is negative there. To the right of $q = a$ marginal revenue is greater than marginal cost, so marginal profit is positive there. At $q = a$ marginal profit changes from negative to positive. This means that profit is decreasing to the left of a and increasing to the right. The point $q = a$ corresponds to a local minimum of profit, and does not maximize profit. It would be a terrible idea for the company to set its production level at $q = a$.

(b) We see from the figure in the problem that just to the left of $q = b$ marginal revenue is greater than marginal cost, so marginal profit is positive there. Just to the right of $q = b$ marginal revenue is less than marginal cost, so marginal profit is negative there. At $q = b$ marginal profit changes from positive to negative. This means that profit is increasing to the left of b and decreasing to the right. The point $q = b$ corresponds to a local maximum of profit. In fact, since the area between the MC and MR curves in the figure in the text between $q = a$ and $q = b$ is bigger than the area between $q = 0$ and $q = a$, $q = b$ is in fact a global maximum.

Solutions for Section 4.5

1. (a) (i) The average cost of quantity q is given by the formula $C(q)/q$. So average cost at $q = 25$ is given by $C(25)/25$. From the graph, we see that $C(25) \approx 200$, so $a(q) \approx 200/25 \approx \$8$ per unit. To interpret this graphically, note that $a(q) = C(q)/q = (C(q) - 0)/(q - 0)$. This is the formula for the slope of a line from the origin to a point $(q, C(q))$ on the curve. So $a(25)$ is the slope of a line connecting $(0, 0)$ to $(25, C(25))$. See Figure 4.18.

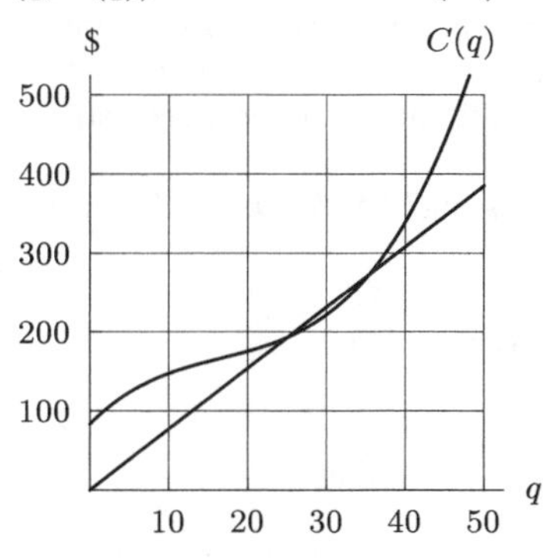

Figure 4.18

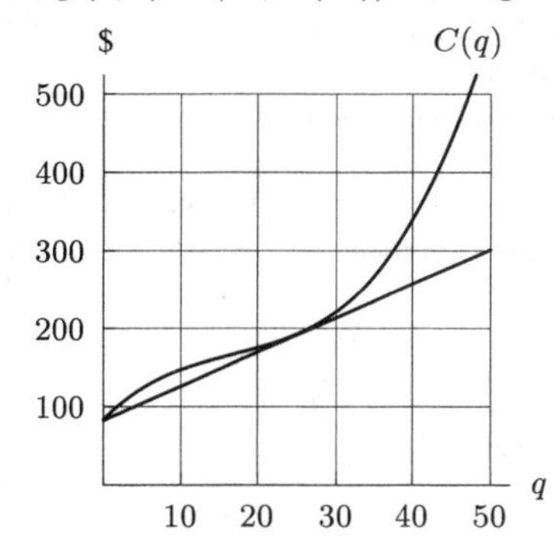

Figure 4.19

(ii) The marginal cost is $C'(q)$. This derivative is the slope of the tangent line to $C(q)$ at $q = 25$. To estimate this slope, we draw the tangent line, shown in Figure 4.19. From this plot, we see that the points $(50, 300)$ and $(0, 100)$ are approximately on this line, so its slope is approximately $(300 - 100)/(50 - 0) = 4$. Thus, $C'(25) \approx \$4$ per unit.

(b) We know that $a(q)$ is minimized where $a(q) = C'(q)$. Using the graphical interpretations from parts (i) and (ii), we see that $a(q)$ is minimized where the line passing from $(q, C(q))$ to the origin is also tangent to the curve. To find such points, a variety of lines passing through the origin and the curve are shown in Figure 4.20. The line which is also a tangent touches the curve at $q \approx 30$. So $q \approx 30$ units minimizes $a(q)$.

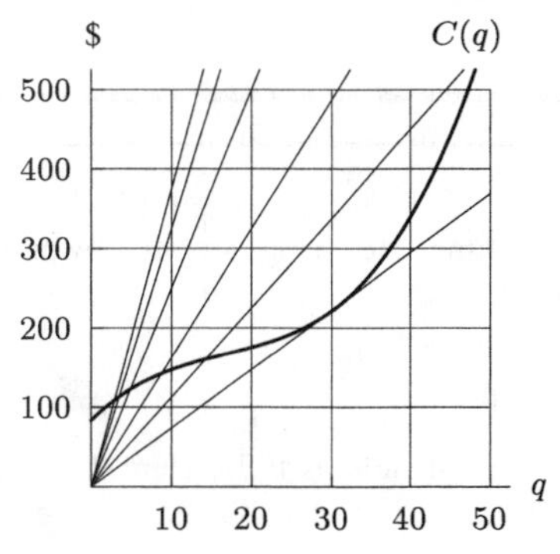

Figure 4.20

5. **(a)** The marginal cost tells us that additional units produced would cost about $10 each, which is below the average cost, so producing them would reduce average cost.

(b) It is impossible to determine the effect on profit from the information given. Profit depends on both cost and revenue, $\pi = R - C$, but we have no information on revenue.

9. The graph of the average cost function is shown in Figure 4.21.

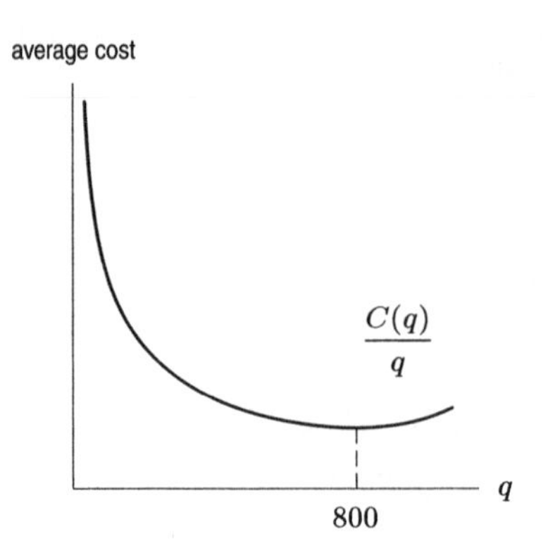

Figure 4.21

13. **(a)** Since the graph is concave down, the average cost gets smaller as q increases. This is because the cost per item gets smaller as q increases. There is no value of q for which the average cost is minimized since for any q_0 larger than q the average cost at q_0 is less than the average cost at q. Graphically, the average cost at q is the slope of the line going through the origin and through the point $(q, C(q))$. Figure 4.22 shows how as q gets larger, the average cost decreases.

(b) The average cost will be minimized at some q for which the line through $(0, 0)$ and $(q, c(q))$ is tangent to the cost curve. This point is shown in Figure 4.23.

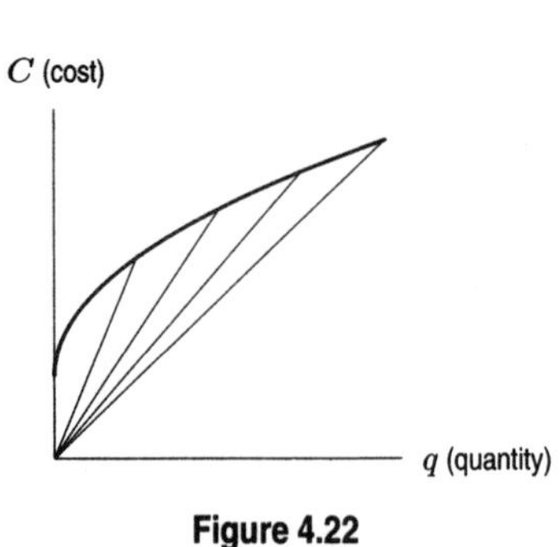

Figure 4.22

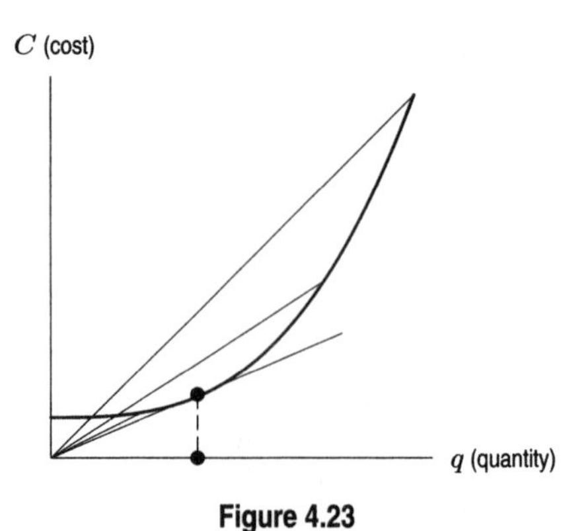

Figure 4.23

Solutions for Section 4.6

1. The effect on demand is approximately E times the change in price. A price increase cause a decrease in demand and a price decrease cause an increase in demand.

(a) Demand decreases by about $2(3\%) = 6\%$.

(b) Demand increases by about $2(3\%) = 6\%$.

5. Table 4.6 of Section 4.6 gives the elasticity of milk as 0.31. Here $0 \leq E \leq 1$, so the demand is inelastic, meaning that changes in price will not change the demand so much. This is expected for milk; as a staple food item, people will continue to buy it even if the price goes up.

9. **(a)** There were good substitutes for slaves in city occupations - including free blacks. There were no good substitutes in the countryside.

(b) They were from the countryside, where there was no satisfactory substitute for slaves.

13. **(a)** We use

$$E = \left|\frac{p}{q}\cdot\frac{dq}{dp}\right|.$$

So we approximate dq/dp at $p = 1.00$.

$$\frac{dq}{dp} \approx \frac{2440-2765}{1.25-1.00} = \frac{-325}{0.25} = -1300$$

$$\text{so} \quad E = \left|\frac{p}{q}\cdot\frac{dq}{dp}\right| \approx \left|\frac{1.00}{2765}\cdot(-1300)\right| = 0.470$$

Since $E = 0.470 < 1$, demand for the candy is inelastic at $p = 1.00$.

(b) At $p = 1.25$,

$$\frac{dq}{dp} \approx \frac{1980-2440}{1.50-1.25} = \frac{-460}{0.25} = -1840$$

$$\text{so} \quad E = \left|\frac{p}{q}\cdot\frac{dq}{dp}\right| \approx \left|\frac{1.25}{2440}\cdot(-1840)\right| = 0.943$$

At $p = 1.5$,

$$\frac{dq}{dp} \approx \frac{1660-1980}{1.75-1.50} = \frac{-320}{0.25} = -1280$$

$$\text{so} \quad E = \left|\frac{p}{q}\cdot\frac{dq}{dp}\right| \approx \left|\frac{1.50}{1980}\cdot(-1280)\right| = 0.970$$

At $p = 1.75$,

$$\frac{dq}{dp} \approx \frac{1175-1660}{2.00-1.75} = \frac{-485}{0.25} = -1940$$

$$\text{so} \quad E = \left|\frac{p}{q}\cdot\frac{dq}{dp}\right| \approx \left|\frac{1.75}{1660}\cdot(-1940)\right| = 2.05$$

At $p = 2.00$,

$$\frac{dq}{dp} \approx \frac{800-1175}{2.25-2.00} = \frac{-375}{0.25} = -1500$$

$$\text{so} \quad E = \left|\frac{p}{q}\cdot\frac{dq}{dp}\right| \approx \left|\frac{2.00}{1175}\cdot(-1500)\right| = 2.55$$

At $p = 2.25$,

$$\frac{dq}{dp} \approx \frac{430-800}{2.50-2.25} = \frac{-370}{0.25} = -1480$$

$$\text{so} \quad E = \left|\frac{p}{q}\cdot\frac{dq}{dp}\right| \approx \left|\frac{2.25}{800}\cdot(-1480)\right| = 4.16$$

Examination of the elasticities for each of the prices suggests that elasticity gets larger as price increases. In other words, at higher prices, an increase in price will cause a larger drop in demand than the same size price increase at a lower price level. This can be explained by the fact that people will not pay too much for candy, as it is somewhat of a "luxury" item.

(c) Elasticity is approximately equal to 1 at $p = \$1.25$ and $p = \$1.50$.

(d)

Table 4.1

p(\$)	q	Revenue $= p\cdot q$ (\$)
1.00	2765	2765
1.25	2440	3050
1.50	1980	2970
1.75	1660	2905
2.00	1175	2350
2.25	800	1800
2.50	430	1075

We can see that revenue is maximized at $p = \$1.25$, with $p = \$1.50$ a close second, which agrees with part (c).

17. Demand is elastic at all prices. No matter what the price is, you can increase revenue by lowering the price. In the end, you would lower your prices all the way to zero. This is not a realistic example, but it is mathematically possible. It would correspond, for instance, to the demand equation $q = 1/p^2$, which gives revenue $R = pq = 1/p$ which is decreasing for all prices $p > 0$.

21. The approximation $E_{\text{cross}} \approx |\frac{\Delta q/q}{\Delta p/p}|$ shows that the cross-price elasticity measures the ratio of the fractional change in quantity of chicken demanded to the fractional change in the price of beef. Thus, for example, a 1% increase in the price of beef will stimulate a E_{cross}% increase in the demand for chicken, presumably because consumers will react to the price rise in beef by switching to chicken. The cross-price elasticity measures the strength of this switch.

Solutions for Section 4.7

1. Substituting $t = 0, 10, 20, \ldots, 70$ into the function $P = 3.9(1.03)^t$ gives the values in Table 4.2. Notice that the agreement is very close, reflecting the fact that an exponential function models the growth well over the period 1790–1860.

Table 4.2 *Predicted versus actual US population 1790–1860, in millions. (exponential model)*

Year	Actual	Predicted	Year	Actual	Predicted
1790	3.9	3.9	1830	12.9	12.7
1800	5.3	5.2	1840	17.1	17.1
1810	7.2	7.0	1850	23.2	23.0
1820	9.6	9.5	1860	31.4	30.9

5. **(a)** If we graph the data, we see that it looks like logistic growth. (See Figure 4.24.) But logistic growth also makes sense from a common-sense viewpoint. As VCRs "catch on," the percentage of households which have them will at first grow exponentially but then slow down after more and more people have them. Eventually, nearly everyone who will buy one already has and the percentage levels off.

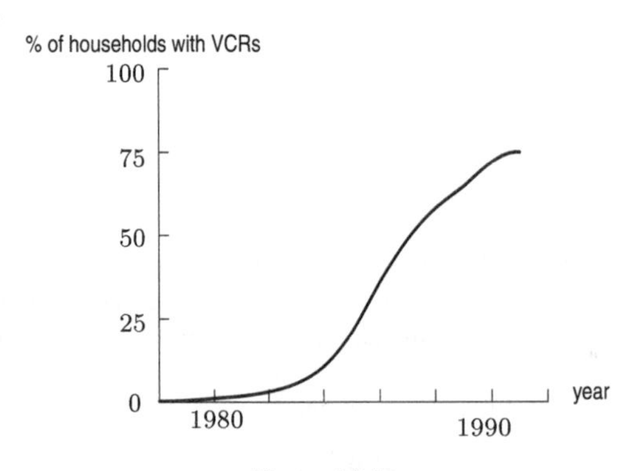

Figure 4.24

(b) The point of diminishing returns happens around 36%. This predicts a carrying capacity of 70% which is pretty close to the 71.9% we see in 1990 and 1991.

(c) 75%

(d) The limiting value predicts the percentage of households with a television set that will eventually have VCRs. This model predicts that there will never be a time when more than 75% of the households with TVs also have VCRs, which seems reasonable.

9. **(a)** We use $k = 1.78$ as a rough approximation. We let $L = 5000$ since the problem tells us that 5000 people eventually get the virus. This means the limiting value is 5000.

(b) We know that

$$P(t) = \frac{5000}{1 + Ce^{-1.78t}} \quad \text{and} \quad P(0) = 10$$

so

$$\begin{aligned} 10 &= \frac{5000}{1 + Ce^0} = \frac{5000}{1 + C} \\ 10(1 + C) &= 5000 \\ 1 + C &= 500 \\ C &= 499. \end{aligned}$$

(c) We have $P(t) = \dfrac{5000}{1 + 499e^{-1.78t}}$. This function is graphed in Figure 4.25.

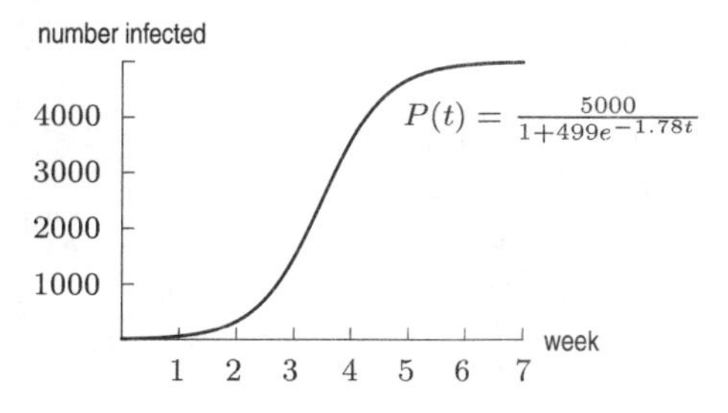

Figure 4.25

(d) The point of diminishing returns appears to be at the point $(3.5, 2500)$; that is, after 3 and a half weeks and when 2500 people are infected.

13. (a) The dose-response curve for product C crosses the minimum desired response line last, so it requires the largest dose to achieve the desired response. The dose-response curve for product B crosses the minimum desired response line first, so it requires the smallest dose to achieve the desired response.

(b) The dose-response curve for product A levels off at the highest point, so it has the largest maximum response. The dose-response curve for product B levels off at the lowest point, so it has the smallest maximum response.

(c) Product C is the safest to administer because its slope in the safe and effective region is the least, so there is a broad range of dosages for which the drug is both safe and effective.

17. When 50 mg of the drug is administered, it is effective for 85 percent of the patients and lethal for 6 percent.

Solutions for Section 4.8

1.

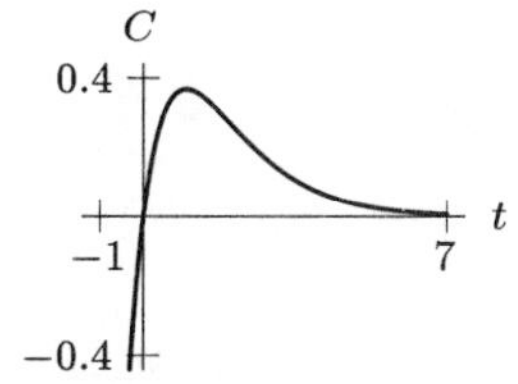

Figure 4.26: $a = 1$

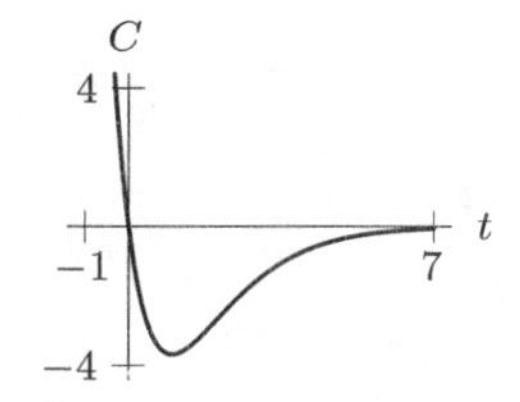

Figure 4.27: $a = -10$

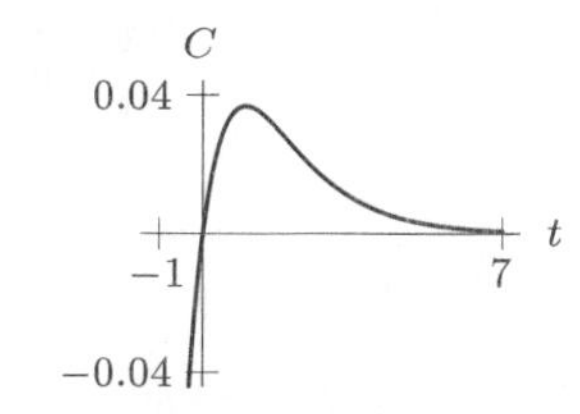

Figure 4.28: $a = 0.1$

The parameter a apparently affects the height and direction of $C = ate^{-bt}$. If a is positive, the "hump" is above the t-axis. If a is negative, it's below the t-axis. If $a > 0$, the larger the value of a, the larger the maximum value of C. If $a < 0$, the more negative the value of a, the smaller the minimum value of C.

5. For cigarettes the peak concentration is approximately 17 ng/ml, the time until peak concentration is about 10 minutes, and the nicotine is at first eliminated quickly, but after about 45 minutes it is eliminated more slowly.

For chewing tobacco, the peak concentration is approximately 14 ng/ml and the time until peak concentration is about 30 minutes (when chewing stops). The nicotine is eliminated at a slow, somewhat erratic rate.

For nicotine gum the peak concentration is about 10 ng/ml and the time until peak concentration is approximately 45 minutes. The nicotine is eliminated at a very slow but steady rate.

9. **(a)** Varying the a parameter changes the peak concentration proportionally, but does not change the time to reach peak concentration .
(b) The value of the peak concentration decreases as b increases and the time until peak concentration decreases as b increases.
(c) As a gets larger, the time until peak concentration decreases because it is only affected by the constant in the exponent. The value of the peak concentration is not affected as a changes because the peak always occurs at $t = 1/b$ and if $a = b$, the value of the peak concentration is $c = a \cdot \frac{1}{a} e^{-a \cdot 1/a} = e^{-1}$, independent of a.

Solutions for Chapter 4 Review

1.

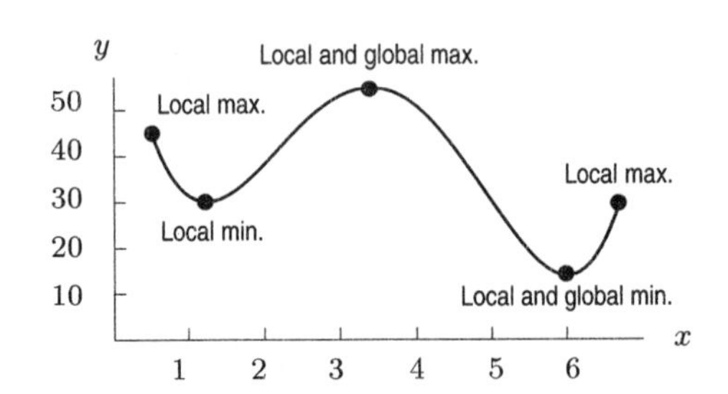

5. **(a)** Differentiating $f(x) = x + \sin x$ produces $f'(x) = 1 + \cos x$. A second differentiation produces $f''(x) = -\sin x$.
(b) $f'(x)$ is defined for all x and $f'(x) = 0$ when $x = \pi$. Thus π is the critical point of f.
(c) $f''(x)$ is defined for all x and $f''(x) = 0$ when $x = 0$, $x = \pi$ and $x = 2\pi$. Since the concavity of f changes at each of these points they are all inflection points.
(d) $f(0) = 0$, $f(2\pi) = 2\pi$, $f(\pi) = \pi$. So f has a global minimum at $x = 0$ and a global maximum at $x = 2\pi$.
(e) Plotting the function $f(x)$ for $0 \le x \le 2\pi$ gives the graph shown in Figure 4.29:

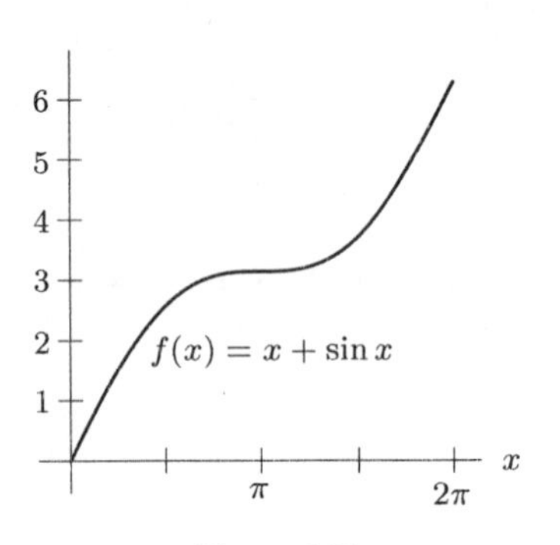

Figure 4.29

9. **(a)** Increasing for $x > 0$, decreasing for $x < 0$.
(b) $f(0)$ is a local and global minimum, and f has no global maximum.

13. We know that the maximum (or minimum) profit can occur when

$$\text{Marginal cost} = \text{Marginal revenue} \quad \text{or} \quad MC = MR.$$

From the table it appears that $MC = MR$ at $q \approx 2500$ and $q \approx 4500$. To decide which one corresponds to the maximum profit, look at the marginal profit at these points. Since

$$\text{Marginal profit} = \text{Marginal revenue} - \text{Marginal cost}$$

(or $M\pi = MR - MC$), we compute marginal profit at the different values of q in Table 4.3:

Table 4.3

q	1000	2000	3000	4000	5000	6000
$M\pi = MR - MC$	-22	-4	4	7	-5	-22

From the table, at $q \approx 2500$, we see that profit changes from decreasing to increasing, so $q \approx 2500$ gives a local minimum. At $q \approx 4500$, profit changes from increasing to decreasing, so $q \approx 4500$ is a local maximum. See Figure 4.30. Therefore, the global maximum occurs at $q = 4500$ or at the endpoint $q = 1000$.

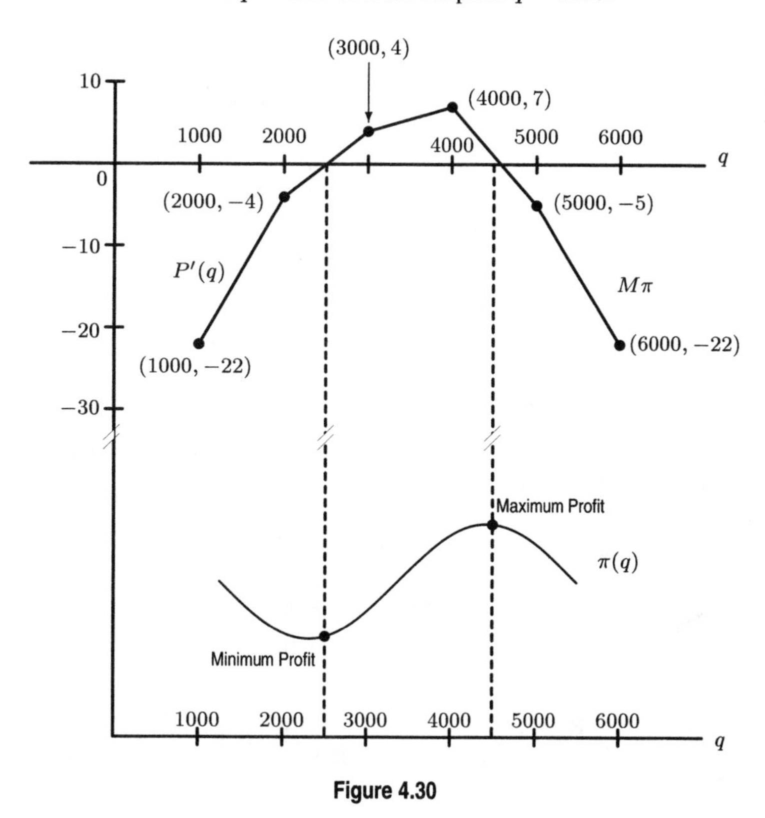

Figure 4.30

17. **(a)** Profit $= \pi = R - C$; profit is maximized when the slopes of the two graphs are equal, at around $q = 350$. See Figure 4.31.

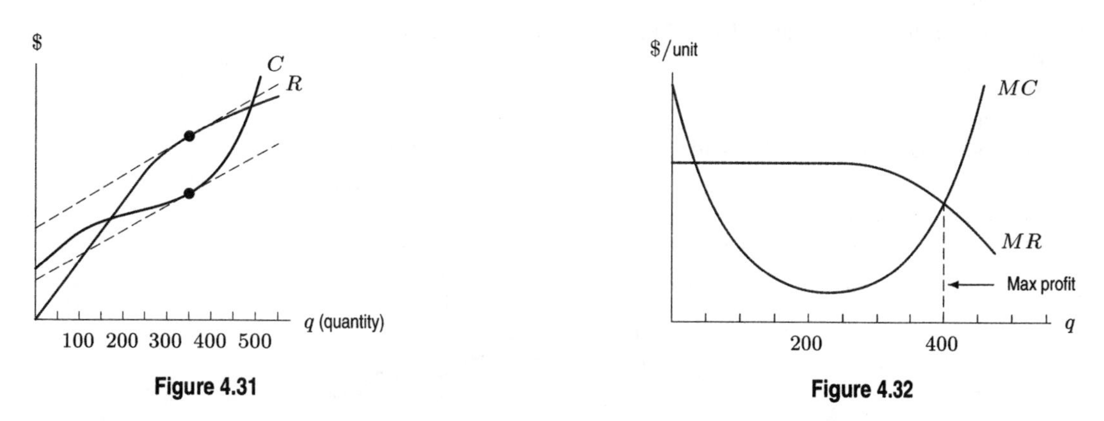

Figure 4.31

Figure 4.32

(b) The graphs of MR and MC are the derivatives of the graphs of R and C. Both R and C are increasing everywhere, so MR and MC are everywhere positive. The cost function is concave down and then concave up, so MC is decreasing and then increasing. The revenue function is linear and then concave down, so MR is constant and then decreasing. See Figure 4.32.

21. The point F representing the fixed costs is the vertical intercept of the cost function, $C(q)$.

The break-even level of production is where $R(q) = C(q)$, so the point B is where the two graphs intersect. There are two such points in Figure 4.33. We have marked the one at which production first becomes profitable.

The marginal cost is $C'(q)$, so the point M is where the slope of $C(q)$ is minimum.

The average cost is minimized when $a(q) = C'(q)$. Since $a(q)$ equals the slope of a line joining the origin to a point $(q, C(q))$ on the graph of $C(q)$, the ponit A is where such a line is tangent to the graph of $C(q)$.

The profit is $P(q) = R(q) - C(q)$, so the point P occurs where the two graphs are furthest apart and $R(q) > C(q)$. See Figure 4.33.

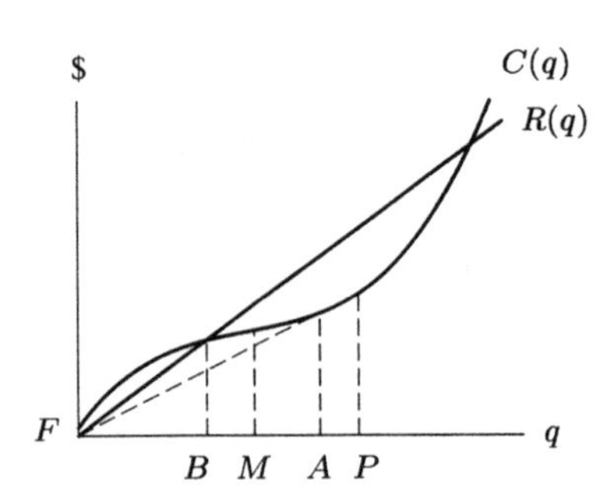

Figure 4.33

25. Marginal cost equals $C'(q) = k$. On the other hand, by the preceding exercise, marginal revenue equals $p(1 - 1/E)$. Maximum profit will occur when the two are equal

$$k = p(1 - 1/E)$$

Thus

$$k/p = 1 - 1/E$$
$$1/E = 1 - k/p$$

On the other hand,

$$\frac{\text{Profit}}{\text{Revenue}} = \frac{\text{Revenue} - \text{Cost}}{\text{Revenue}} = 1 - \frac{\text{Cost}}{\text{Revenue}} = 1 - \frac{kq}{pq} = 1 - \frac{k}{p} = \frac{1}{E}$$

29. **(a)** Quadratic polynomial (degree 2) with negative leading coefficient.
(b) Exponential.
(c) Logistic.
(d) Logarithmic.
(e) This is a quadratic polynomial (degree 2) and positive leading coefficient.
(f) Exponential.
(g) Surge
(h) Periodic
(i) This looks like a cubic polynomial (degree 3) and negative leading coefficient.

33. The critical points of f occur where f' is zero. These two points are indicated in the figure below.

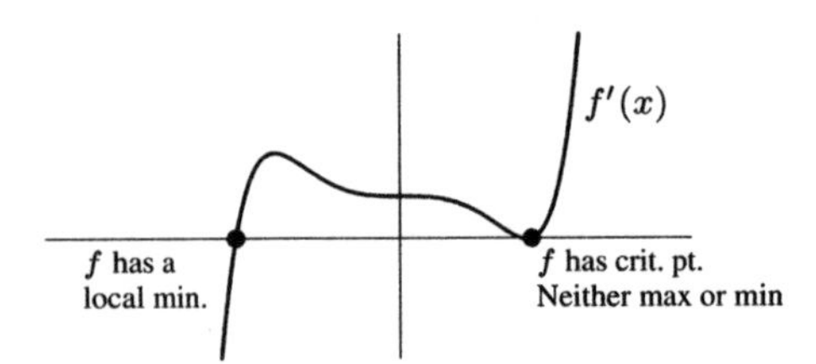

Note that the point labeled as a local minimum of f is not a critical point of f'.

CHAPTER FIVE

Solutions for Section 5.1

1. We use Distance = Rate × Time on each subinterval with $\Delta t = 3$.

$$\text{Underestimate} = 0 \cdot 3 + 10 \cdot 3 + 25 \cdot 3 + 45 \cdot 3 = 240,$$
$$\text{Overestimate} = 10 \cdot 3 + 25 \cdot 3 + 45 \cdot 3 + 75 \cdot 3 = 465.$$

We know that

$$240 \leq \text{Distance traveled} \leq 465.$$

A better estimate is the average. We have

$$\text{Distance traveled} \approx \frac{240 + 465}{2} = 352.5.$$

The car travels about 352.5 feet during these 12 seconds.

5. **(a)** The velocity is 30 miles/hour for the first 2 hours, 40 miles/hour for the next $1/2$ hour, and 20 miles/hour for the last 4 hours. The entire trip lasts $2 + 1/2 + 4 = 6.5$ hours, so we need a scale on our horizontal (time) axis running from 0 to 6.5. Between $t = 0$ and $t = 2$, the velocity is constant at 30 miles/hour, so the velocity graph is a horizontal line at 30. Likewise, between $t = 2$ and $t = 2.5$, the velocity graph is a horizontal line at 40, and between $t = 2.5$ and $t = 6.5$, the velocity graph is a horizontal line at 20. The graph is shown in Figure 5.1.

(b) How can we visualize distance traveled on the velocity graph given in Figure 5.1? The velocity graph looks like the top edges of three rectangles. The distance traveled on the first leg of the journey is (30 miles/hour)(2 hours), which is the height times the width of the first rectangle in the velocity graph. The distance traveled on the first leg of the trip is equal to the area of the first rectangle. Likewise, the distances traveled during the second and third legs of the trip are equal to the areas of the second and third rectangles in the velocity graph. It appears that distance traveled is equal to the area under the velocity graph.

In Figure 5.2, the area under the velocity graph in Figure 5.1 is shaded. Since this area is three rectangles and the area of each rectangle is given by Height × Width, we have

$$\begin{aligned}\text{Total area} &= (30)(2) + (40)(1/2) + (20)(4)\\ &= 60 + 20 + 80 = 160.\end{aligned}$$

The area under the velocity graph is equal to distance traveled.

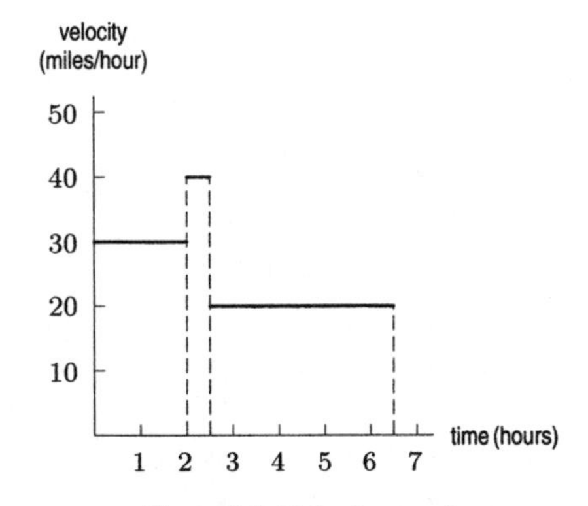

Figure 5.1: Velocity graph

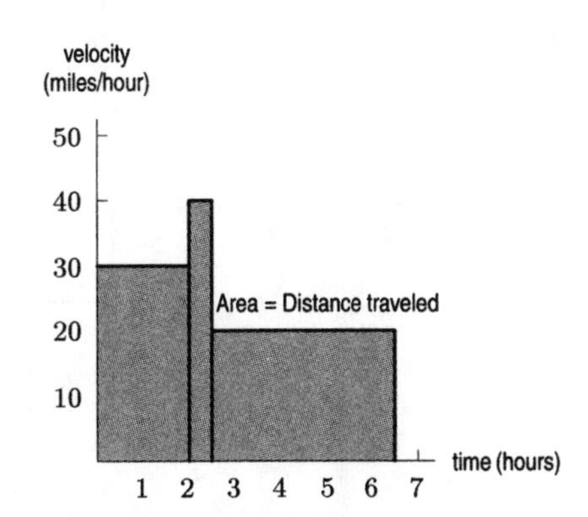

Figure 5.2: The area under the velocity graph gives distance traveled

9. The table gives the rate of oil consumption in billions of barrels per year. To find the total consumption, we use left-hand and right-hand Riemann sums. We have

$$\text{Left-hand sum} = (22.3)(5) + (21.3)(5) + (23.9)(5) + (24.9)(5) = 462.0 \text{ bn barrels.}$$
$$\text{Right-hand sum} = (21.3)(5) + (23.9)(5) + (24.9)(5) + (27.0)(5) = 485.5 \text{ bn barrels.}$$
$$\text{Average of left- and right-hand sums} = \frac{462.0 + 485.5}{2} = 473.75 \text{ bn barrels.}$$

The consumption of oil between 1980 and 2000 is about 474 billion barrels.

13. **(a)** Note that 15 minutes equals 0.25 hours. Lower estimate $= 11(0.25) + 10(0.25) = 5.25$ miles. Upper estimate $= 12(0.25) + 11(0.25) = 5.75$ miles.

(b) Lower estimate $= 11(0.25) + 10(0.25) + 10(0.25) + 8(0.25) + 7(0.25) + 0(0.25) = 11.5$ miles. Upper estimate $= 12(0.25) + 11(0.25) + 10(0.25) + 10(0.25) + 8(0.25) + 7(0.25) = 14.5$ miles.

17. The rate at which the fish population grows varies between 10 fish per month and about 22 fish per month. If the rate of change were constant at the lower bound of 10 during the entire 12-month period, we obtain an underestimate for the total change in the fish population of $(10)(12) = 120$ fish. An overestimate is $(22)(12) = 264$ fish. The actual increase in the fish population is equal to the area under the curve. (Notice that the units of the area are height units times width units, or fish per month times months, as we want.) We estimate that the area is about 11 grid squares. The area of each grid square represents an increase of (5 fish per month) $\cdot$ (4 months) $= 20$ fish. We have

$$\text{Total change in fish population} = \text{Area under curve} \approx 11 \cdot 20 = 220.$$

We estimate that the fish population grew by 220 fish during this 12-month period.

Solutions for Section 5.2

1.

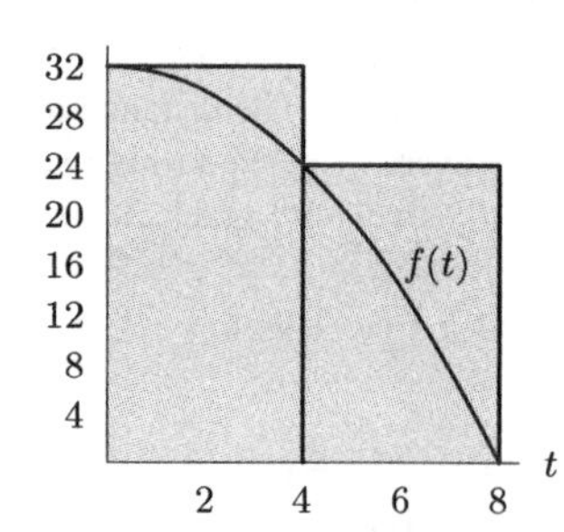

Figure 5.3: Left Sum, $\Delta t = 4$

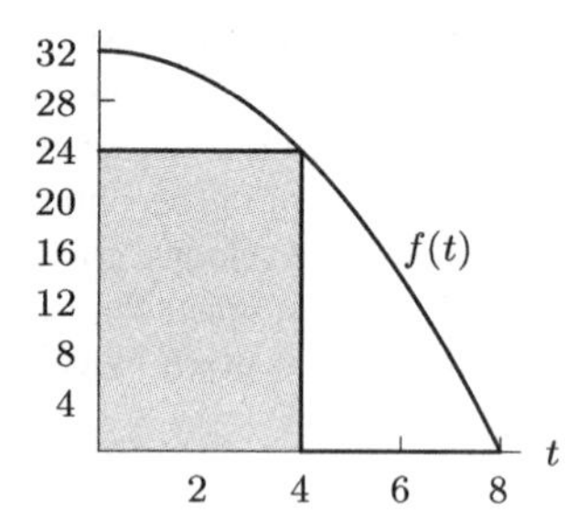

Figure 5.4: Right Sum, $\Delta t = 4$

(a) Left-hand sum $= 32 \cdot 4 + 24 \cdot 4 = 224$.
(b) Right-hand sum $= 24 \cdot 4 + 0 \cdot 4 = 96$.

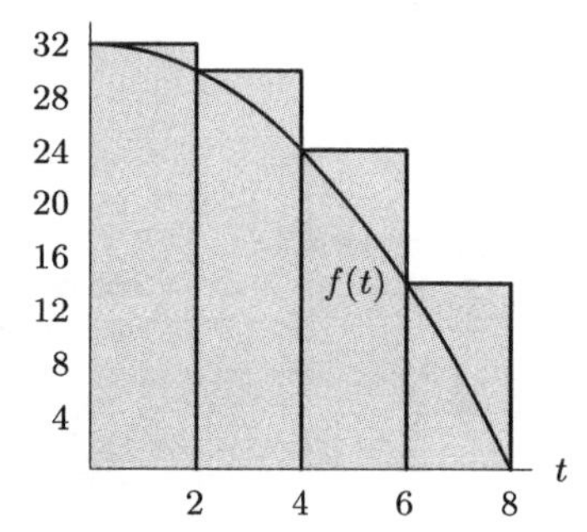

Figure 5.5: Left Sum, $\Delta t = 2$

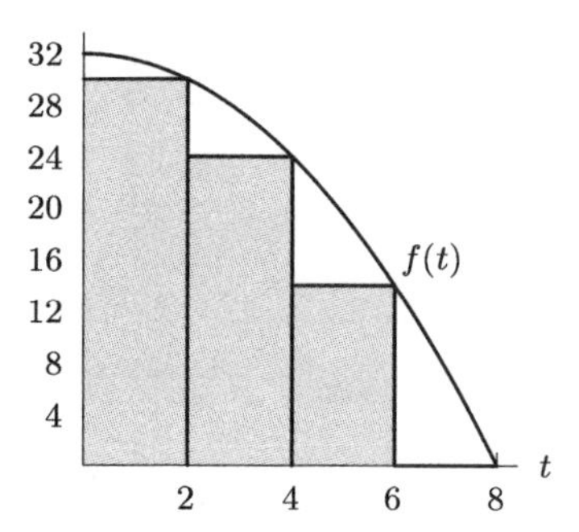

Figure 5.6: Right Sum, $\Delta t = 2$

(c) Left-hand sum $= 32 \cdot 2 + 30 \cdot 2 + 24 \cdot 2 + 14 \cdot 2 = 200$.
(d) Right-hand sum $= 30 \cdot 2 + 24 \cdot 2 + 14 \cdot 2 + 0 \cdot 2 = 136$.

5. **(a)** See Figure 5.7.

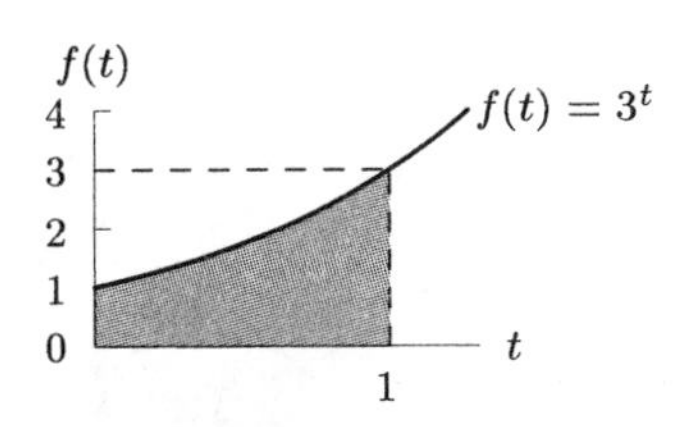

Figure 5.7

The shaded area is approximately 60% of the area of the rectangle 1 unit by 3 units. Therefore,

$$\int_0^1 3^t \, dt \approx 0.60 \cdot 1 \cdot 3 = 1.8.$$

(b) $\int_0^1 3^t \, dt = 1.8205.$

9.

$$\begin{aligned}
\text{Left-hand sum} &= 50 \cdot 3 + 48 \cdot 3 + 44 \cdot 3 + 36 \cdot 3 + 24 \cdot 3 = 606 \\
\text{Right-hand sum} &= 48 \cdot 3 + 44 \cdot 3 + 36 \cdot 3 + 24 \cdot 3 + 8 \cdot 3 = 480 \\
\text{Average} &= \frac{606 + 480}{2} = 543
\end{aligned}$$

So we have $\int_0^{15} f(x)\, dx \approx 543.$

13. To estimate the integral, we count the rectangles under the curve and above the x-axis for the interval $[0, 3]$. There are approximately 17 of these rectangles, and each has area 1, so

$$\int_0^3 f(x)dx \approx 17.$$

17. $\int_1^4 \frac{1}{\sqrt{1+x^2}}\, dx = 1.2$

21. $\int_1^2 (1.03)^t \, dt = 1.0$

25. **(a)** If $\Delta t = 4$, then $n = 2$. We have:

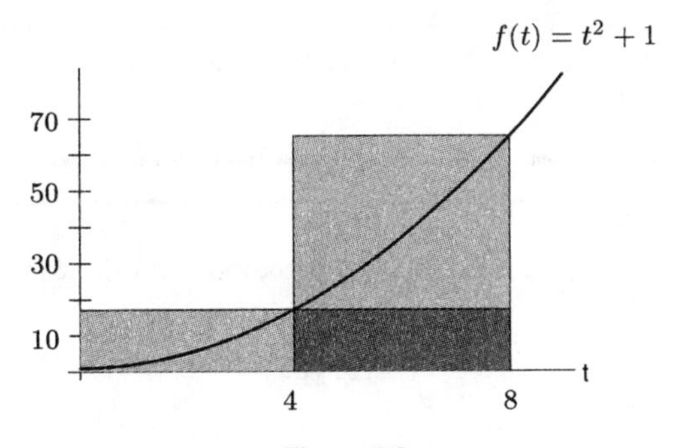

Figure 5.8

$$\begin{aligned}
\text{Underestimate of total change } &= f(0)\Delta t + f(4)\Delta t = 1 \cdot 4 + 17 \cdot 4 = 72. \\
\text{Overestimate of total change } &= f(4)\Delta t + f(8)\Delta t = 17 \cdot 4 + 65 \cdot 4 = 328.
\end{aligned}$$

(b) If $\Delta t = 2$, then $n = 4$. We have:

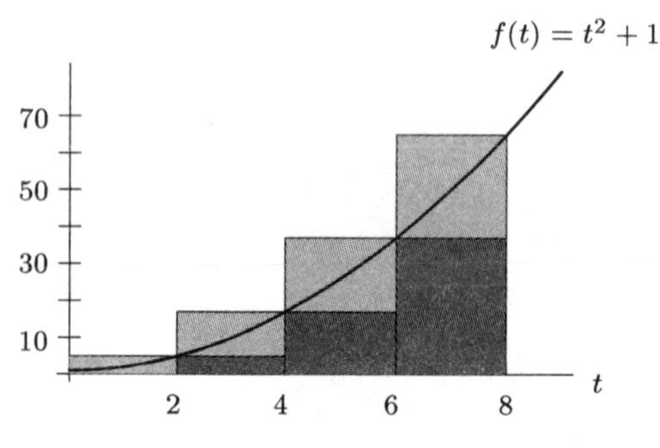

Figure 5.9

$$\begin{aligned}
\text{Underestimate of total change} &= f(0)\Delta t + f(2)\Delta t + f(4)\Delta t + f(6)\Delta t \\
&= 1 \cdot 2 + 5 \cdot 2 + 17 \cdot 2 + 37 \cdot 2 = 120. \\
\text{Overestimate of total change} &= f(2)\Delta t + f(4)\Delta t + f(6)\Delta t + f(8)\Delta t \\
&= 5 \cdot 2 + 17 \cdot 2 + 37 \cdot 2 + 65 \cdot 2 = 248.
\end{aligned}$$

(c) If $\Delta t = 1$, then $n = 8$.

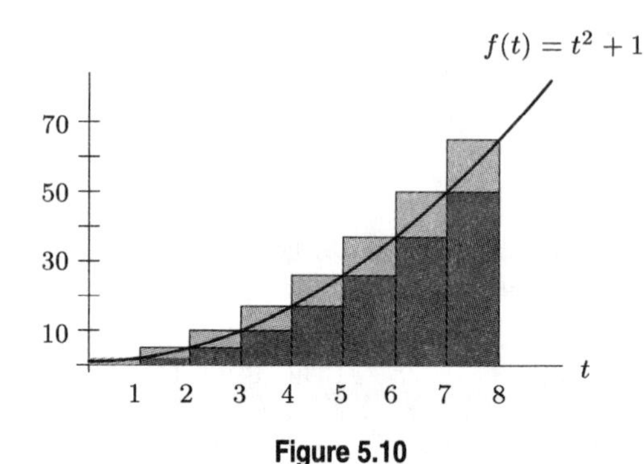

Figure 5.10

Underestimate of total change

$$\begin{aligned}
&= f(0)\Delta t + f(1)\Delta t + f(2)\Delta t + f(3)\Delta t + f(4)\Delta t + f(5)\Delta t + f(6)\Delta t + f(7)\Delta t \\
&= 1 \cdot 1 + 2 \cdot 1 + 5 \cdot 1 + 10 \cdot 1 + 17 \cdot 1 + 26 \cdot 1 + 37 \cdot 1 + 50 \cdot 1 = 148.
\end{aligned}$$

Overestimate of total change

$$\begin{aligned}
&= f(1)\Delta t + f(2)\Delta t + f(3)\Delta t + f(4)\Delta t + f(5)\Delta t + f(6)\Delta t + f(7)\Delta t + f(8)\Delta t \\
&= 2 \cdot 1 + 5 \cdot 1 + 10 \cdot 1 + 17 \cdot 1 + 26 \cdot 1 + 37 \cdot 1 + 50 \cdot 1 + 65 \cdot 1 = 212.
\end{aligned}$$

Solutions for Section 5.3

1. The integral represents the area below the graph of $f(x)$ but above the x-axis. Since each square has area 1, by counting squares and half-squares we find

$$\int_1^6 f(x)\, dx = 8.5.$$

5. The area below the x-axis is greater than the area above the x-axis, so the integral is negative.

9. **(a)** Counting the squares yields an estimate of 16.5, each with area $= 1$, so the total shaded area is approximately 16.5.

(b)

$$\begin{aligned}
\int_0^8 f(x)dx &= (\text{shaded area above } x\text{-axis}) - (\text{shaded area below } x\text{-axis}) \\
&\approx 6.5 - 10 = -3.5
\end{aligned}$$

(c) The answers in (a) and (b) are different because the shaded area below the x-axis is subtracted in order to find the value of the integral in (b).

13. See Figure 5.11.

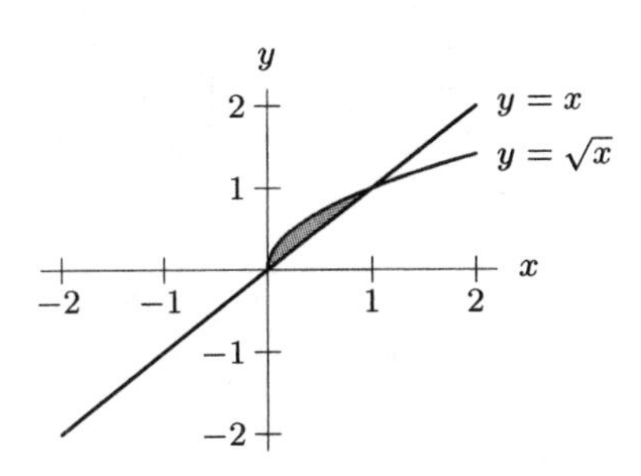

Figure 5.11

Inspection of the graph tell us that the curves intersect at $(0, 0)$ and $(1, 1)$, with $\sqrt{x} \geq x$ for $0 \leq x \leq 1$, so we can find the area by evaluating the integral

$$\int_0^1 (\sqrt{x} - x)dx.$$

Using technology to evaluate the integral, we see

$$\int_0^1 (\sqrt{x} - x)dx \approx 0.1667.$$

So the area between the graphs is about 0.1667.

17. **(a)** The area shaded between 0 and 1 is about the same as the area shaded between -1 and 0, so the area between 0 and 1 is about 0.25. Since this area lies below the x-axis, we estimate that

$$\int_0^1 f(x)\,dx = -0.25.$$

(b) Between -1 and 1, the area above the x-axis approximately equals the area below the x-axis, and so

$$\int_{-1}^1 f(x)\,dx = (0.25) + (-0.25) = 0.$$

(c) We estimate

$$\text{Total area shaded} = 0.25 + 0.25 = 0.5.$$

21. Using a calculator or computer, we find

$$\int_1^4 (x - 3\ln x)\,dx \approx -0.136.$$

The function $f(x) = x - 3\ln x$ crosses the x-axis at $x \approx 1.86$. See Figure 5.12.

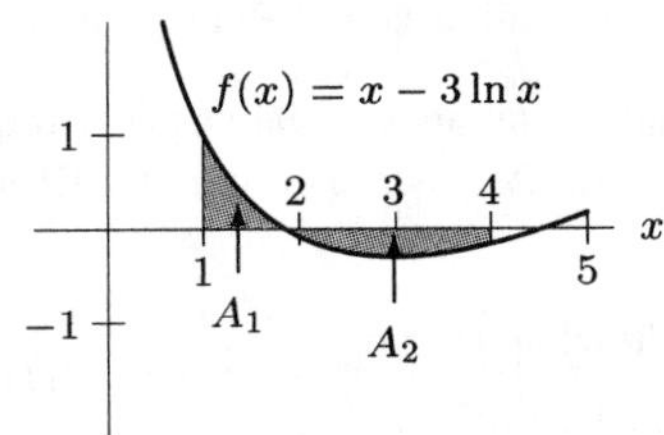

Figure 5.12

We find

$$\int_1^{1.86} (x - 3\ln x)\,dx = \text{Area above axis } \approx 0.347$$

$$\int_{1.86}^4 (x - 3\ln x)\,dx = -\text{Area below axis } \approx -0.483.$$

Thus, $A_1 \approx 0.347$ and $A_2 \approx 0.483$, so

$$\int_1^4 (x - 3\ln x)\,dx = \text{Area above axis}-\text{Area below axis } \approx 0.347 - 0.483 = -0.136.$$

25. The area below the x-axis is bigger than the area above the x-axis, so the integral is negative. The area above the x-axis appears to be about half the area of the rectangle with area $5 \cdot 2 = 10$, so we estimate the area above the x-axis to be approximately 5. The area below the x-axis appears to be about half the area of the rectangle with area $10 \cdot 3 = 30$, so we estimate the area below the x-axis to be approximately 15. See Figure 5.13. The value of the integral is the area above the x-axis minus the area below the x-axis, so we estimate

$$\int_0^5 f(x)dx \approx 5 - 15 = -10.$$

The correct match for this function is I.

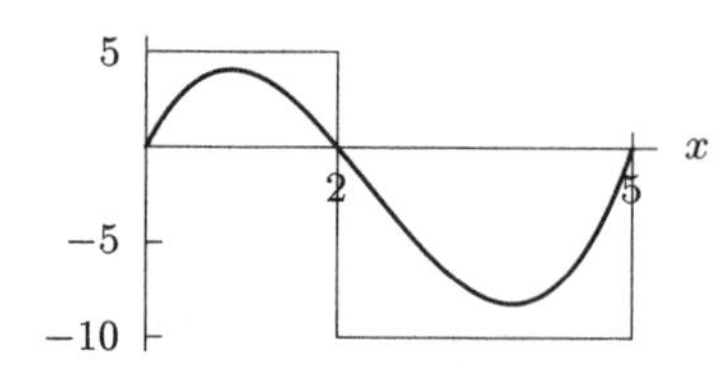

Figure 5.13

29. **(a)** $\int_{-3}^{0} f(x)\,dx = -2.$

(b) $\int_{-3}^{4} f(x)\,dx = \int_{-3}^{0} f(x)\,dx + \int_{0}^{3} f(x)\,dx + \int_{3}^{4} f(x)\,dx = -2 + 2 - \frac{A}{2} = -\frac{A}{2}.$

Solutions for Section 5.4

1. $\int_a^b f(t)dt$ is measured in

$$\left(\frac{\text{miles}}{\text{hours}}\right) \cdot (\text{hours}) = \text{miles}.$$

5. For any t, consider the interval $[t, t + \Delta t]$. During this interval, oil is leaking out at an approximately constant rate of $f(t)$ gallons/minute. Thus, the amount of oil which has leaked out during this interval can be expressed as

$$\text{Amount of oil leaked} = \text{Rate} \times \text{Time} = f(t)\,\Delta t$$

and the units of $f(t)\,\Delta t$ are gallons/minute × minutes = gallons. The total amount of oil leaked is obtained by adding all these amounts between $t = 0$ and $t = 60$. (An hour is 60 minutes.) The sum of all these infinitesimal amounts is the integral

$$\begin{array}{c}\text{Total amount of}\\ \text{oil leaked, in gallons}\end{array} = \int_0^{60} f(t)\,dt.$$

9. Since W is in tons per week and t is in weeks since January 1, 2000, the integral $\int_0^{52} W dt$ gives the amount of waste, in tons, produced during the year 2000.

$$\text{Total waste during the year} = \int_0^{52} 3.75e^{-0.008t}\,dt = 159.5249 \text{ tons}.$$

Since waste removal costs \$15/ton, the cost of waste removal for the company is $159.5249 \cdot 15 = \$2392.87$.

13. **(a)** The area under the curve is greater for species B for the first 5 years. Thus species B has a larger population after 5 years. After 10 years, the area under the graph for species B is still greater so species B has a greater population after 10 years as well.

(b) Unless something happens that we cannot predict now, species A will have a larger population after 20 years. It looks like species A will continue to quickly increase, while species B will add only a few new plants each year.

17. You start traveling toward home. After 1 hour, you have traveled about 5 miles. (The distance traveled is the area under the first bump of the curve. The area is about $1/2$ grid square; each grid square represents 10 miles.) You are now about $10 - 5 = 5$ miles from home. You then decide to turn around. You ride for an hour and a half away from home, in which time you travel around 13 miles. (The area under the next part of the curve is about 1.3 grid squares below the axis.) You are now $5 + 13 = 18$ miles from home and you decide to turn around again at $t = 2.5$. The area above the curve from $t = 2.5$ to $t = 3.5$ has area 18 miles. This means that you ride for another hour, at which time you reach your house ($t = 3.5$.) You still want to ride, so you keep going past your house for a half hour. You make about 8 miles in this time and you decide to stop. So, at the end of the ride, you end up 8 miles away from home.

21. The total number of "worker-hours" is equal to the area under the curve. The total area is about 14.5 boxes. Since each box represents (10 workers)(8 hours) = 80 worker-hours, the total area is 1160 worker-hours. At \$10 per hour, the total cost is \$11,600.

Solutions for Section 5.5

1. The units for the integral $\int_{800}^{900} C'(q)dq$ are $\left(\frac{\text{dollars}}{\text{tons}}\right) \cdot (\text{tons}) = \text{dollars}$.

$\int_{800}^{900} C'(q)dq$ represents the cost of increasing production from 800 tons to 900 tons.

5. **(a)** Total variable cost in producing 400 units is

$$\int_0^{400} C'(q)\,dq.$$

We estimate this integral:

$$\begin{aligned}\text{Left–hand sum} &= 25(100) + 20(100) + 18(100) + 22(100) = 8500;\\ \text{Right–hand sum} &= 20(100) + 18(100) + 22(100) + 28(100) = 8800;\end{aligned}$$

and so $\displaystyle\int_0^{400} C'(q)\,dq \approx \frac{8500 + 8800}{2} = \$8650.$

$$\begin{aligned}\text{Total cost} &= \text{Fixed cost} + \text{Variable cost}\\ &= \$10{,}000 + \$8650 = \$18{,}650.\end{aligned}$$

(b) $C'(400) = 28$, so we would expect that the 401st unit would cost an extra \$28.

9. Using the Fundamental Theorem of Calculus with $f(x) = 4 - x^2$ and $a = 0$, we see that

$$F(b) - F(0) = \int_0^b F'(x)dx = \int_0^b (4 - x^2)dx.$$

We know that $F(0) = 0$, so

$$F(b) = \int_0^b (4 - x^2)dx.$$

Using Riemann sums to estimate the integral for values of b, we get

Table 5.1

b	0.0	0.5	1.0	1.5	2.0	2.5
$F(b)$	0	1.958	3.667	4.875	5.333	4.792

Solutions for Chapter 5 Review

1. (a) Let's begin by graphing the data given in the table; see Figure 5.14. The total amount of pollution entering the lake during the 30-day period is equal to the shaded area. The shaded area is roughly 40% of the rectangle measuring 30 units by 35 units. Therefore, the shaded area measures about $(0.40)(30)(35) = 40$ units. Since the units are kilograms, we estimate that 420 kg of pollution have entered the lake.

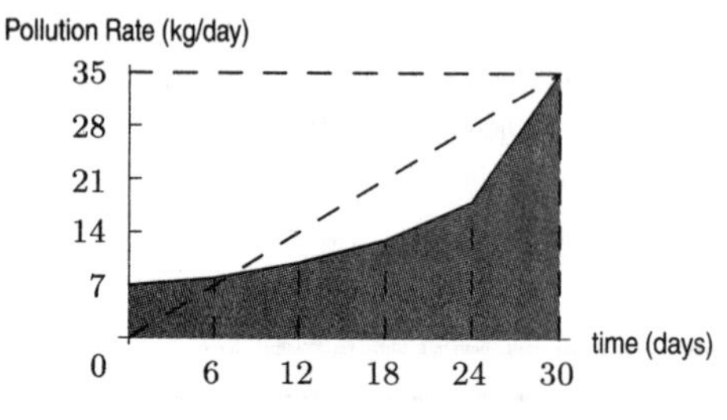

Figure 5.14

(b) Using left and right sums, we have

$$\text{Underestimate} = (7)(6) + (8)(6) + (10)(6) + (13)(6) + (18)(6) = 336 \text{ kg}.$$

$$\text{Overestimate} = (8)(6) + (10)(6) + (13)(6) + (18)(6) + (35)(6) = 504 \text{ kg}.$$

5. $\int_0^{10} 2^{-x}\, dx = 1.44$

9. $\int_2^3 \frac{-1}{(r+1)^2}\, dr = -0.083$

13. Looking at the area under the graph we see that after 5 years Tree B is taller while Tree A is taller after 10 years.

17. In Figure 5.15 the area A_1 is largest, A_2 is next, and A_3 is smallest. We have

$$\text{I} = \int_a^b f(x)\, dx = A_1, \quad \text{II} = \int_a^c f(x)\, dx = A_1 - A_2, \quad \text{III} = \int_a^e f(x)\, dx = A_1 - A_2 + A_3,$$

$$\text{IV} = \int_b^e f(x)\, dx = -A_2 + A_3, \quad \text{V} = \int_b^c f(x)\, dx = -A_2.$$

The relative sizes of A_1, A_2, and A_3 mean that I is positive and largest, III is next largest (since $-A_2 + A_3$ is negative, but less negative than $-A_2$), II is next largest, but still positive (since A_1 is larger than A_2). The integrals IV and V are both negative, but V is more negative. Thus

$$\text{V} < \text{IV} < 0 < \text{II} < \text{III} < \text{I}.$$

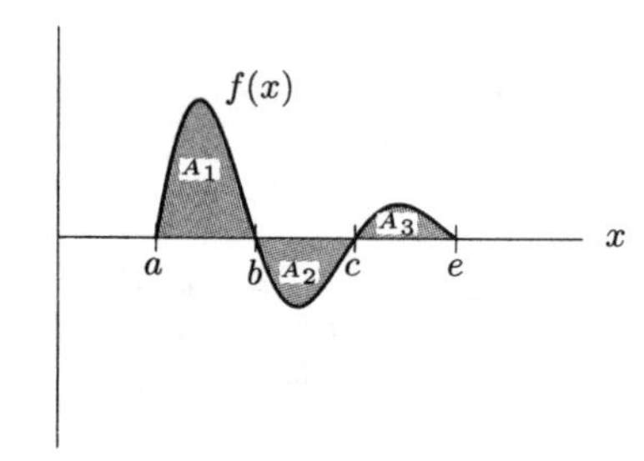

Figure 5.15

21. From $t = 0$ to $t = 3$, you are moving away from home ($v > 0$); thereafter you move back toward home. So you are the farthest from home at $t = 3$. To find how far you are then, we can measure the area under the v curve as about 9 squares, or $9 \cdot 10$ km/hr $\cdot$ 1 hr $= 90$ km. To find how far away from home you are at $t = 5$, we measure the area from $t = 3$ to $t = 5$ as about 25 km, except that this distance is directed toward home, giving a total distance from home during the trip of $90 - 25 = 65$ km.

25. **(a)**

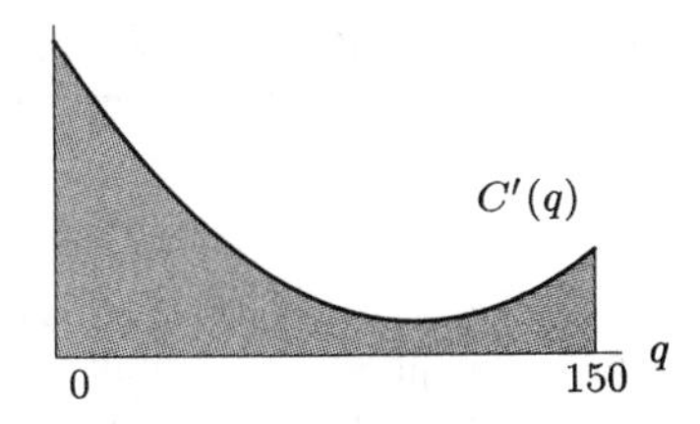

Figure 5.16

The total variable cost of producing 150 units is represented by the area under the graph of $C'(q)$ between 0 and 150, or

$$\int_0^{150} (0.005q^2 - q + 56)dq.$$

(b) An estimate of the total cost of producing 150 units is given by

$$20{,}000 + \int_0^{150} (0.005q^2 - q + 56)dq.$$

This represents the fixed cost ($20,000) plus the variable cost of producing 150 units, which is represented by the integral. Using a calculator, we see

$$\int_0^{150} (0.005q^2 - q + 56)dq \approx 2{,}775.$$

So the total cost is approximately

$$\$20{,}000 + \$2{,}775 = \$22{,}775.$$

(c) $C'(150) = 0.005(150)^2 - 150 + 56 = 18.5$. This means that the marginal cost of the 150th item is 18.5. In other words, the 151st item will cost approximately \$18.50.

(d) $C(151)$ is the total cost of producing 151 items. This can be found by adding the total cost of producing 150 items (found in part (b)) and the additional cost of producing the 151st item ($C'(150)$, found in (c)). So we have

$$C(151) \approx 22{,}775 + 18.50 = \$22{,}793.50.$$

29. **(a)** The distance traveled is the integral of the velocity, so in T seconds you fall

$$\int_0^T 49(1 - 0.8187^t)\,dt.$$

(b) We want the number T for which

$$\int_0^T 49(1 - 0.8187^t)\,dt = 5000.$$

We can use a calculator or computer to experiment with different values for T, and we find $T \approx 107$ seconds.

33. **(a)** The mouse changes direction (when its velocity is zero) at about times 17, 23, and 27.

(b) The mouse is moving most rapidly to the right at time 10 and most rapidly to the left at time 40.

(c) The mouse is farthest to the right when the integral of the velocity, $\int_0^t v(t)\,dt$, is most positive. Since the integral is the sum of the areas above the axis minus the areas below the axis, the integral is largest when the velocity is zero at about 17 seconds. The mouse is farthest to the left of center when the integral is most negative at 40 seconds.

(d) The mouse's speed decreases during seconds 10 to 17, from 20 to 23 seconds, and from 24 seconds to 27 seconds.

(e) The mouse is at the center of the tunnel at any time t for which the integral from 0 to t of the velocity is zero. This is true at time 0 and again somewhere around 35 seconds.

Solutions to Problems on the Second Fundamental Theorem of Calculus

1. By the Second Fundamental Theorem, $G'(x) = x^3$.

5. **(a)** If $F(b) = \int_0^b 2^x\,dx$ then $F(0) = \int_0^0 2^x\,dx = 0$ since we are calculating the area under the graph of $f(x) = 2^x$ on the interval $0 \le x \le 0$, or on no interval at all.

(b) Since $f(x) = 2^x$ is always positive, the value of F will increase as b increases. That is, as b grows larger and larger, the area under $f(x)$ on the interval from 0 to b will also grow larger.

(c) Using a calculator or a computer, we get

$$F(1) = \int_0^1 2^x\,dx \approx 1.4,$$
$$F(2) = \int_0^2 2^x\,dx \approx 4.3,$$
$$F(3) = \int_0^3 2^x\,dx \approx 10.1.$$

9. Note that $\int_a^b (g(x))^2\,dx = \int_a^b (g(t))^2\,dt$. Thus, we have

$$\int_a^b \left((f(x))^2 - (g(x))^2\right)\,dx = \int_a^b (f(x))^2\,dx - \int_a^b (g(x))^2\,dx = 12 - 3 = 9.$$

CHAPTER SIX

Solutions for Section 6.1

1. By counting grid squares, we find $\int_1^6 f(x)\,dx = 8.5$, so the average value of f is $\frac{8.5}{6-1} = \frac{8.5}{5} = 1.7$.

5. Average value $= \frac{1}{10-0}\int_0^{10} e^t\,dt \approx 2202.55$

9. **(a)** The average inventory is given by the formula

$$\frac{1}{90-0}\int_0^{90} 5000(0.9)^t\,dt.$$

Using a calculator yields

$$\int_0^{90} 5000(0.9)^t\,dt \approx 47452.5$$

so the average inventory is

$$\frac{47452.5}{90} \approx 527.25.$$

(b) The function is graphed in Figure 6.1. The area of a rectangle of height 527.25 is equal to the area under the curve.

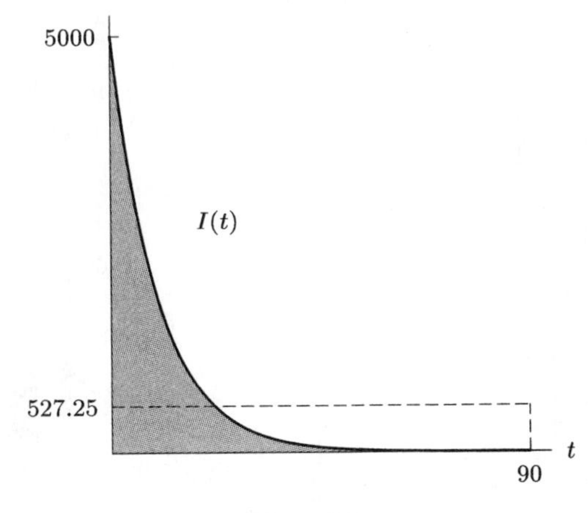

Figure 6.1

13. **(a)** $Q(10) = 4(0.96)^{10} \approx 2.7$. $Q(20) = 4(0.96)^{20} \approx 1.8$

(b)

$$\frac{Q(10)+Q(20)}{2} \approx 2.21$$

(c) The average value of Q over the interval is about 2.18.

(d) Because the graph of Q is concave up between $t = 10$ and $t = 20$, the area under the curve is less than what is obtained by connecting the endpoints with a straight line.

17. We'll show that in terms of the average value of f,

$$\text{I} > \text{II} = \text{IV} > \text{III}$$

Using the definition of average value and the fact that f is even, we have

$$\begin{aligned}\text{Average value of } f \text{ on II} &= \frac{\int_0^2 f(x)dx}{2} = \frac{\frac{1}{2}\int_{-2}^2 f(x)dx}{2} \\ &= \frac{\int_{-2}^2 f(x)dx}{4} \\ &= \text{Average value of } f \text{ on IV.}\end{aligned}$$

Since f is decreasing on [0,5], the average value of f on the interval $[0, c]$, where $0 \leq c \leq 5$, is decreasing as a function of c. The larger the interval the more low values of f are included. Hence

$$\begin{array}{c}\text{Average value of } f \\ \text{on } [0,1]\end{array} > \begin{array}{c}\text{Average value of } f \\ \text{on } [0,2]\end{array} > \begin{array}{c}\text{Average value of } f \\ \text{on } [0,5]\end{array}$$

Solutions for Section 6.2

1. **(a)** Looking at the figure in the problem we see that the equilibrium price is roughly \$30 giving an equilibrium quantity of 125 units.

   (b) Consumer surplus is the area above p^* and below the demand curve. Graphically this is represented by the shaded area in Figure 6.2. From the graph we can estimate the shaded area to be roughly 14 squares where each square represents (\$25/unit)·(10 units). Thus the consumer surplus is approximately

$$14 \cdot \$250 = \$3500.$$

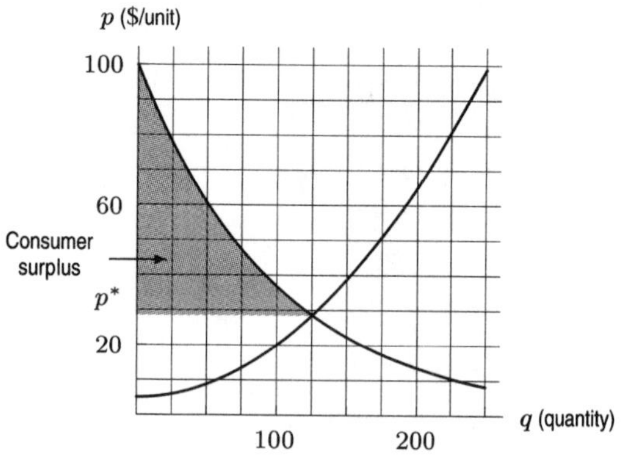

Figure 6.2

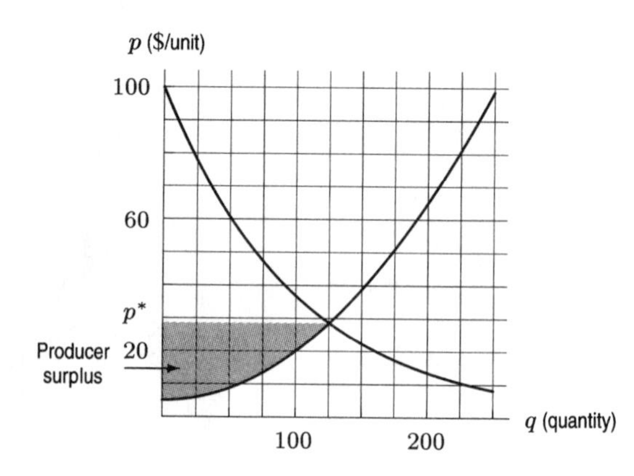

Figure 6.3

Producer surplus is the area under p^* and above the supply curve. Graphically this is represented by the shaded area in Figure 6.3. From the graph we can estimate the shaded area to be roughly 8 squares where each square represents (\$25/unit)·(10 units). Thus the producer surplus is approximately

$$8 \cdot \$250 = \$2000$$

   (c) We have

$$\begin{aligned}\text{Total gains from trade} &= \text{Consumer surplus} + \text{producer surplus} \\ &= \$3500 + \$2000 \\ &= \$5500.\end{aligned}$$

5. **(a)** In Table 6.1, the quantity q increases as the price p decreases, while in Table 6.2, q increases as p increases. Therefore, the demand data is in Table 6.1 and the supply data is in Table 6.2.

(b) It appears that the equilibrium price is $p^* = 25$ dollars per unit and the equilibrium quantity is $q^* = 400$ units sold at this price.

(c) To estimate the consumer surplus, we use the demand data in Table 6.1. We use a Riemann sum using the price from the demand data minus the equilibrium price of 25.

$$\text{Left sum} = (60-25)\cdot 100 + (50-25)\cdot 100 + (41-25)\cdot 100 + (32-25)\cdot 100 = 8300.$$

$$\text{Right sum} = (50-25)\cdot 100 + (41-25)\cdot 100 + (32-25)\cdot 100 + (25-25)\cdot 100 = 4800.$$

We average the two to estimate that

$$\text{Consumer surplus} \approx \frac{8300+4800}{2} = 6550.$$

To estimate the producer surplus, we use the supply data in Table 6.2. We use a Riemann sum using the equilibrium price of 25 minus the price from the supply data.

$$\text{Left sum} = (25-10)\cdot 100 + (25-14)\cdot 100 + (25-18)\cdot 100 + (25-22)\cdot 100 = 3600.$$

$$\text{Right sum} = (25-14)\cdot 100 + (25-18)\cdot 100 + (25-22)\cdot 100 + (25-25)\cdot 100 = 2100.$$

We average the two to estimate that

$$\text{Producer surplus} \approx \frac{3600+2100}{2} = 2850.$$

9. The total gains from trade at the equilibrium price is shaded in Figure 6.4. We see in Figures 6.5 and 6.6 that if the price is artificially high or low, the quantity sold is less than q^*. Thus, the total gains from trade are reduced.

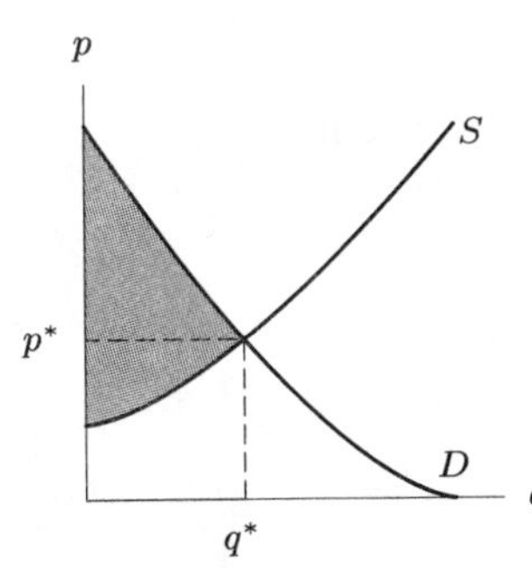

Figure 6.4: Shade area: Total gains from trade at equilibrium price

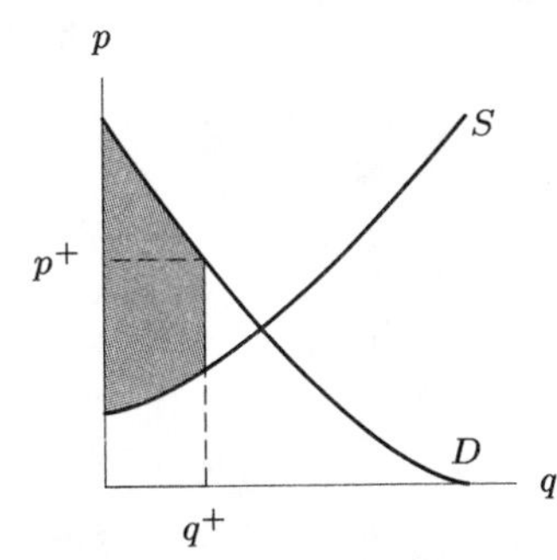

Figure 6.5: Shade area: Gains when price is artificially high

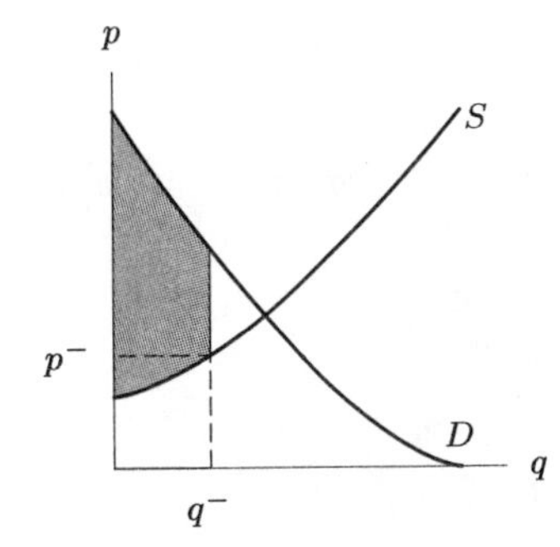

Figure 6.6: Shade area: Gains when price is artificially low

13. **(a)** $p^*q^* =$ the total amount paid for q^* of the good at equilibrium.

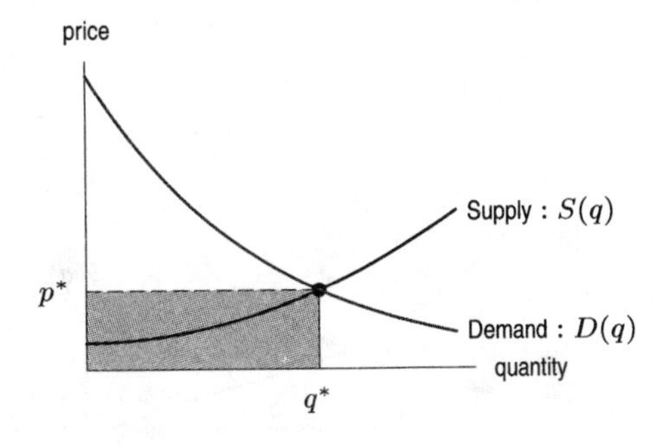

(b) $\int_0^{q^*} D(q)\,dq =$ the maximum consumers would be willing to pay if they had to pay the highest price acceptable to them for each additional unit of the good.

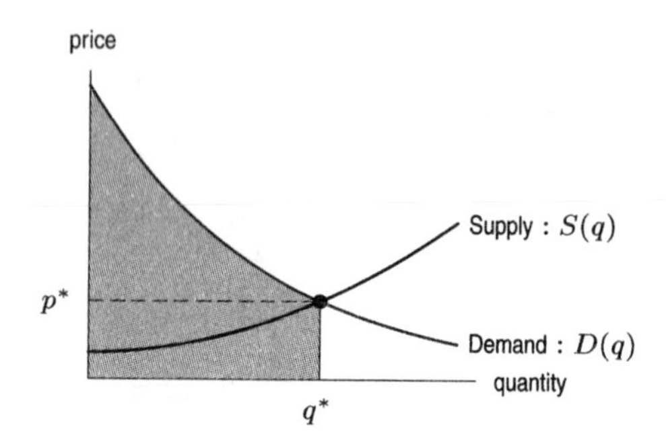

(c) $\int_0^{q^*} S(q)\,dq =$ the minimum suppliers would be willing to accept if they were paid the minimum price acceptable to them for each additional unit of the good.

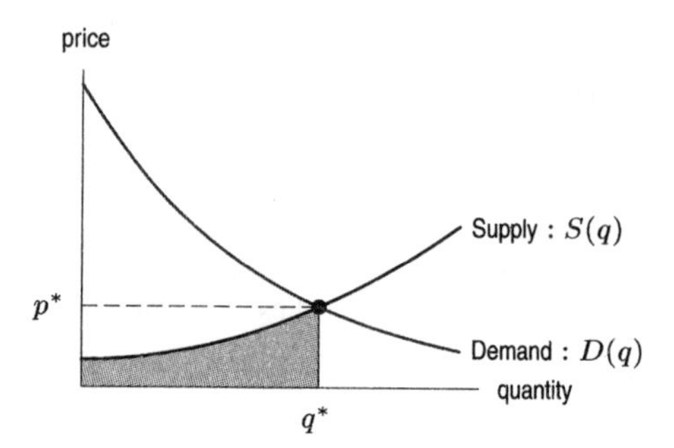

(d) $\int_0^{q^*} D(q)\,dq - p^*q^* =$ consumer surplus.

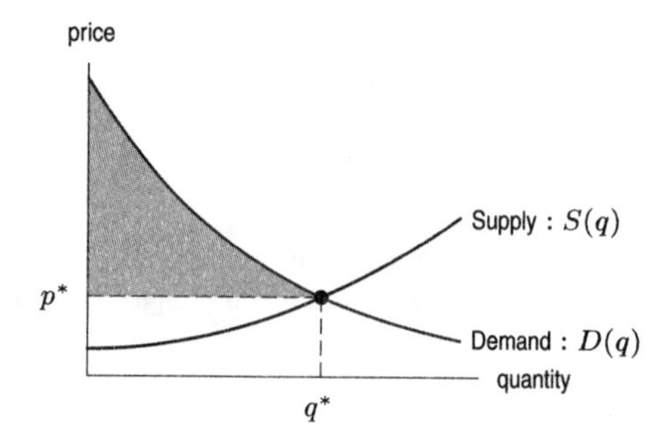

(e) $p^*q^* - \int_0^{q^*} S(q)\,dq =$ producer surplus.

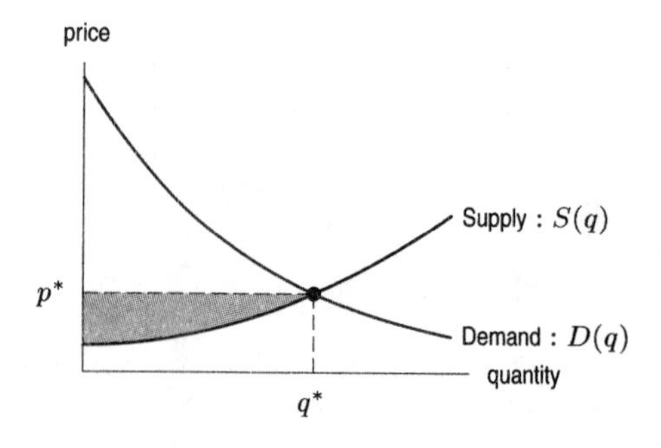

(f) $\int_0^{q^*} (D(q) - S(q))\, dq$ = producer surplus and consumer surplus.

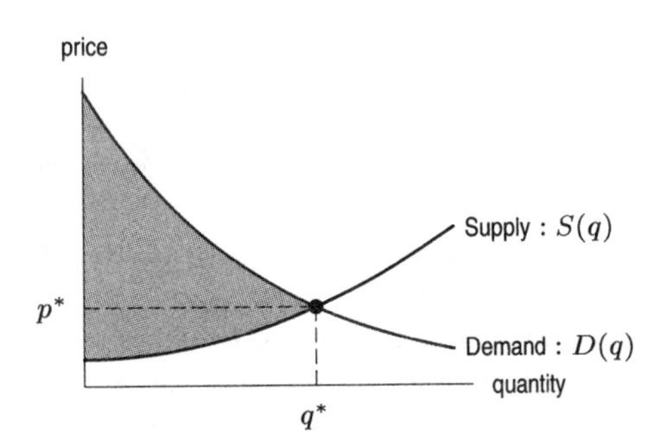

Solutions for Section 6.3

1.

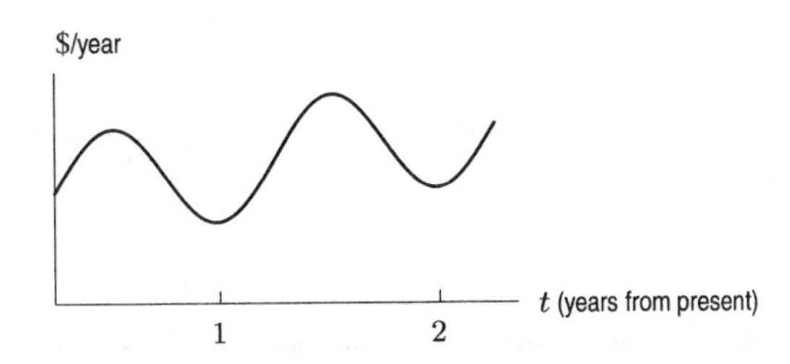

The graph reaches a peak each summer, and a trough each winter. The graph shows sunscreen sales increasing from cycle to cycle. This gradual increase may be due in part to inflation and to population growth.

5. (a) We first find the present value, P, of the income stream:

$$P = \int_0^{10} 6000e^{-0.05t}\, dt = \$47{,}216.32.$$

We use the present value to find the future value, F:

$$F = Pe^{rt} = 47126.32e^{0.05(10)} = \$77{,}846.55.$$

(b) The income stream contributed \$6000 per year for 10 years, or \$60,000. The interest earned was $77{,}846.55 - 60{,}000 = \$17{,}846.55$.

9. (a) Since the income stream is \$7035 million per year and the interest rate is 8.5%,

$$\begin{aligned}\text{Present value} &= \int_0^{1} 7035e^{-0.085t}\, dt \\ &= 6744.31 \text{ million dollars.}\end{aligned}$$

The present value of Intel's profits over a one-year time period is about 6744 million dollars.

(b) The value at the end of one year is $6744.31e^{0.085(1)} = 7342.64$, or about 7343, million dollars. This is the value, at the end of one year, of Intel's profits over a one-year time period.

13. We want to find the value of T making the present value of income stream (\$80,000/year) equal to \$130,000. Thus we want to find the time T at which

$$130{,}000 = \int_0^{T} 80{,}000e^{-0.085t}\, dt.$$

Trying a few values of T, we get $T \approx 1.75$. It takes approximately one year and nine months for the present value of the profit generated by the new machinery to equal the cost of the machinery.

Solutions for Section 6.4

1. Between 1995 and 1996, absolute increase $= 10.6 - 8.2 = 2.4$ million. Between 1999 and 2000, absolute increase $= 21.8 - 18.8 = 3.0$ million.

   Between 1995 and 1996, relative increase $= 2.4/8.2 = 29.3\%$. Between 1999 and 2000, relative increase $= 3.0/18.8 = 16.0\%$.

5. The relative birth rate is $\frac{30}{1000} = 0.03$ and the relative death rate is $\frac{20}{1000} = 0.02$. The relative rate of growth is $0.03 - 0.02 = 0.01$

9. The area between the graph of the relative rate of change and the t-axis is 0 (because the areas above and below are equal). Since the change in $\ln P(t)$ is the area under the curve

$$\begin{aligned} \ln P(10) - \ln P(10) &= \int_0^{10} \frac{P'(t)}{P(t)}\,dt = 0 \\ \ln\left(\frac{P(10)}{P(0)}\right) &= 0 \\ \frac{P(10)}{P(0)} &= e^0 = 1 \\ P(10) &= P(0). \end{aligned}$$

   The population at $t = 0$ and at $t = 10$ are equal.

13. Since the relative rate of change is positive for $0 \le t < 5$ and negative for $5 < t \le 10$, the function f is increasing for $0 \le t < 5$ and decreasing for $5 < t \le 10$.

Solutions for Chapter 6 Review

1. The average value equals

$$\frac{1}{3}\int_0^3 f(x)\,dx = \frac{24}{3} = 8.$$

5. **(a)** At the end of one hour $t = 60$, and $H = 22°$C.

   (b)

$$\begin{aligned} \text{Average temperature} &= \frac{1}{60}\int_0^{60} (20 + 980e^{-0.1t})dt \\ &= \frac{1}{60}(10976) = 183°\text{C}. \end{aligned}$$

   (c) Average temperature at beginning and end of hour $= (1000 + 22)/2 = 511°$C. The average found in part (b) is smaller than the average of these two temperatures because the bar cools quickly at first and so spends less time at high temperatures. Alternatively, the graph of H against t is concave up.

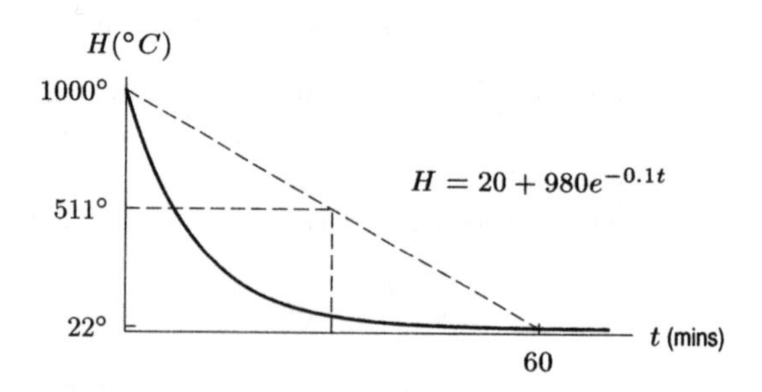

9. It appears that the area under a line at about $y = 17$ is approximately the same as the area under $f(x)$ on the interval $x = a$ to $x = b$, so we estimate that the average value is about 17. See Figure 6.7.

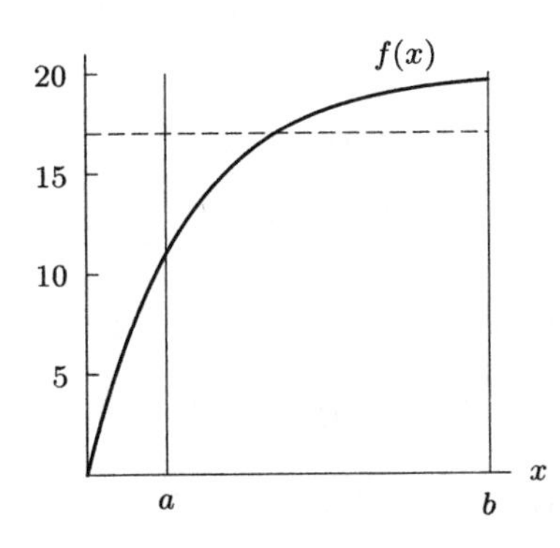

Figure 6.7

13. **(a)** The equilibrium price is \$30 per unit, and the equilibrium quantity is 6000.

(b) The region representing the consumer surplus is the shaded triangle in Figure 6.8 with area $\frac{1}{2} \cdot 6000 \cdot 70 = 210{,}000$. The consumer surplus is \$210,000.

The area representing the producer surplus, shaded in Figure 6.9, is about 7 grid squares, each of area 10,000. The producer surplus is about \$70,000.

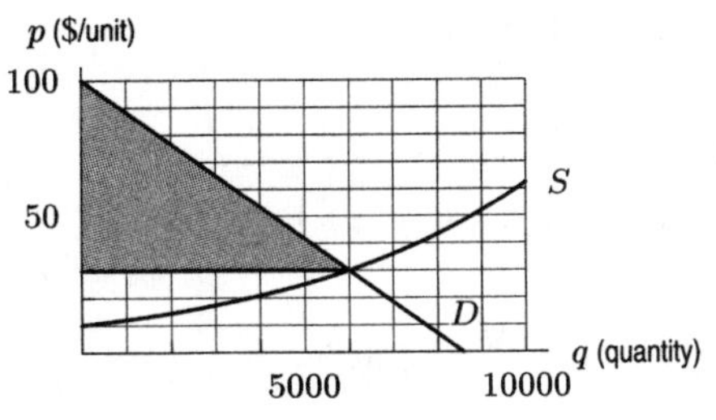

Figure 6.8

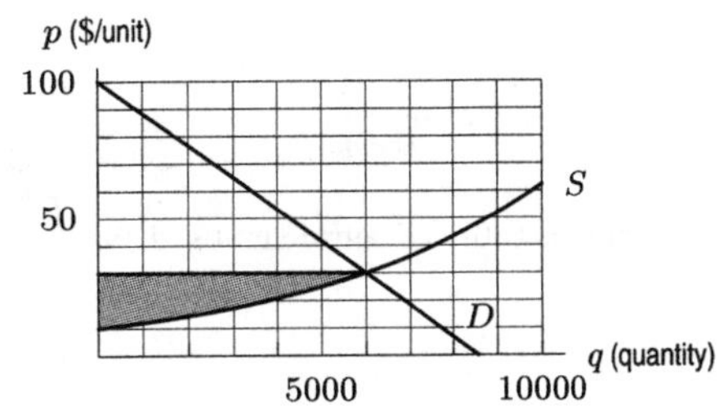

Figure 6.9

17. We first find the present value:

$$\text{Present value} = \int_0^{20} 12000e^{-0.06t}\,dt = \$139{,}761.16.$$

We use the present value to find the future value:

$$\text{Future value} = Pe^{rt} = 139761.16e^{0.06(20)} = \$464{,}023.39.$$

21. At any time t, the company receives income of $s(t) = 50e^{-t}$ thousands of dollars per year. Thus the present value is

$$\begin{aligned}\text{Present value} &= \int_0^2 s(t)e^{-0.06t}\,dt \\ &= \int_0^2 (50e^{-t})e^{-0.06t}\,dt \\ &= \$41{,}508.\end{aligned}$$

25. **(a)** The population of Nicaragua is growing exponentially and we use $P = P_0 a^t$. In 1990 (at $t = 0$), we know $P = 3.6$, so $P_0 = 3.6$. Nicaragua is growing at a rate of 3.4% per year, so $a = 1.034$. We have

$$P = 3.6(1.034)^t.$$

(b) We first find the average rate of change between $t = 0$ (1990) and $t = 1$ (1991). At $t = 0$, we know $P = 3.6$ million. At $t = 1$, we find that $P = 3.6(1.034)^t = 3.7224$ million. We have

$$\text{Average rate of change} = \frac{P(1) - P(0)}{1 - 0} = \frac{3.7224 - 3.6}{1} = 0.1224 \text{ people/yr.}$$

Between 1990 and 1991, the population of Nicaragua grew by 0.1224 million people per year, or 122,400 people per year.

When $t = 2$ (1992), we have $P = 3.6(1.034)^2 = 3.8490$. Between 1991 and 1992, we have

$$\text{Average rate of change} = \frac{P(2) - P(1)}{2 - 1} = \frac{3.8490 - 3.7224}{1} = 0.1266 \text{ people/yr.}$$

Between 1990 and 1991, the population of Nicaragua grew by 0.1266 million people per year, or 126,600 people per year. Notice that the population increase was greater between 1991 and 1992. This is because the population was greater in 1991 than in 1990 and so there were more people to have babies.

(c) The relative rate of change between 1990 and 1991 is the absolute rate of change divided by the population in 1990, which is $0.1224/3.6 = 0.034$. The relative rate of change is 3.4% per year, as we expected. Between 1991 and 1992, the relative rate of change is $0.1266/3.7224 = 0.034$. Over small intervals, the relative rate of change will always be approximately 3.4% per year.

29. The rate at which petroleum is being used t years after 1990 is given by

$$r(t) = 1.4 \cdot 10^{20}(1.02)^t \text{ joules/year.}$$

Between 1990 and M years later

$$\begin{aligned}\text{Total quantity of petroleum used} &= \int_0^M 1.4 \cdot 10^{20}(1.02)^t\, dt = 1.4 \cdot 10^{20} \left.\frac{(1.02)^t}{\ln(1.02)}\right|_0^M \\ &= \frac{1.4 \cdot 10^{20}}{\ln(1.02)}((1.02)^M - 1) \text{ joules.}\end{aligned}$$

Setting the total quantity used equal to 10^{22} gives

$$\begin{aligned}\frac{1.4 \cdot 10^{20}}{\ln(1.02)}\left((1.02)^M - 1\right) &= 10^{22} \\ (1.02)^M &= \frac{100\ln(1.02)}{1.4} + 1 = 2.41 \\ M &= \frac{\ln(2.41)}{\ln(1.02)} \approx 45 \text{ years.}\end{aligned}$$

So we will run out of petroleum in 2035.

CHAPTER SEVEN

Solutions for Section 7.1

1. $5x$
5. $\frac{x^5}{5}$.
9. $\frac{2}{3}z^{\frac{3}{2}}$
13. $-\frac{1}{2z^2}$
17. $\sin t$
21. $f(x) = \frac{1}{4}x$, so $F(x) = \frac{x^2}{8} + C$. $F(0) = 0$ implies that $\frac{1}{8} \cdot 0^2 + C = 0$, so $C = 0$. Thus $F(x) = x^2/8$ is the only possibility.
25. $\frac{3x^2}{2} + C$
29. $\frac{t^{13}}{13} + C$.
33. $5e^z + C$
37. $\frac{x^3}{3} + 2x^2 - 5x + C$
41. $\frac{x^2}{2} + 2x^{1/2} + C$
45. $-\cos t + C$
49. $5\sin x + 3\cos x + C$
53. $10x - 4\cos(2x) + C$
57. An antiderivative is $F(x) = 2e^{3x} + C$. Since $F(0) = 5$, we have $5 = 2e^0 + C = 2 + C$, so $C = 3$. The answer is $F(x) = 2e^{3x} + 3$.

Solutions for Section 7.2

1. We use the substitution $w = x^2 + 1$, $dw = 2xdx$.
$$\int 2x(x^2+1)^5 dx = \int w^5 dw = \frac{w^6}{6} + C = \frac{1}{6}(x^2+1)^6 + C.$$
Check: $\frac{d}{dx}(\frac{1}{6}(x^2+1)^6 + C) = 2x(x^2+1)^5$.
5. We use the substitution $w = x^2 + 1$, $dw = 2xdx$.
$$\int \frac{2x}{\sqrt{x^2+1}} dx = \int w^{-1/2} dw = 2w^{1/2} + C = 2\sqrt{x^2+1} + C.$$
Check: $\frac{d}{dx}(2\sqrt{x^2+1} + C) = \frac{2x}{\sqrt{x^2+1}}$.
9. In this case, it seems easier not to substitute.
$$\int y^2(1+y)^2\, dy = \int y^2(y^2+2y+1)\, dy = \int (y^4 + 2y^3 + y^2)\, dy$$
$$= \frac{y^5}{5} + \frac{y^4}{2} + \frac{y^3}{3} + C.$$
Check: $\frac{d}{dy}\left(\frac{y^5}{5} + \frac{y^4}{2} + \frac{y^3}{3} + C\right) = y^4 + 2y^3 + y^2 = y^2(y+1)^2$.

13. We use the substitution $w = 4 - x$, $dw = -dx$.

$$\int \frac{1}{\sqrt{4-x}}\,dx = -\int \frac{1}{\sqrt{w}}\,dw = -2\sqrt{w} + C = -2\sqrt{4-x} + C.$$

Check: $\frac{d}{dx}(-2\sqrt{4-x} + C) = -2 \cdot \frac{1}{2} \cdot \frac{1}{\sqrt{4-x}} \cdot -1 = \frac{1}{\sqrt{4-x}}.$

17. We use the substitution $w = \cos\theta + 5$, $dw = -\sin\theta\,d\theta$.

$$\int \sin\theta(\cos\theta+5)^7\,d\theta = -\int w^7\,dw = -\frac{1}{8}w^8 + C$$
$$= -\frac{1}{8}(\cos\theta+5)^8 + C.$$

Check:

$$\frac{d}{d\theta}\left[-\frac{1}{8}(\cos\theta+5)^8 + C\right] = -\frac{1}{8}\cdot 8(\cos\theta+5)^7\cdot(-\sin\theta)$$
$$= \sin\theta(\cos\theta+5)^7$$

21. We use the substitution $w = x^3+1$, $dw = 3x^2\,dx$, to get

$$\int x^2 e^{x^3+1}\,dx = \frac{1}{3}\int e^w\,dw = \frac{1}{3}e^w + C = \frac{1}{3}e^{x^3+1} + C.$$

Check: $\frac{d}{dx}\left(\frac{1}{3}e^{x^3+1} + C\right) = \frac{1}{3}e^{x^3+1}\cdot 3x^2 = x^2e^{x^3+1}.$

25. We use the substitution $w = 3x - 4$, $dw = 3dx$.

$$\int e^{3x-4}dx = \frac{1}{3}\int e^w dw = \frac{1}{3}e^w + C = \frac{1}{3}e^{3x-4} + C.$$

Check: $\frac{d}{dx}\left(\frac{1}{3}e^{3x-4} + C\right) = \frac{1}{3}e^{3x-4}\cdot 3 = e^{3x-4}.$

29. We use the substitution $w = \ln z$, $dw = \frac{1}{z}\,dz$.

$$\int \frac{(\ln z)^2}{z}\,dz = \int w^2\,dw = \frac{w^3}{3} + C = \frac{(\ln z)^3}{3} + C.$$

Check: $\frac{d}{dz}\left[\frac{(\ln z)^3}{3} + C\right] = 3\cdot\frac{1}{3}(\ln z)^2\cdot\frac{1}{z} = \frac{(\ln z)^2}{z}.$

33. We use the substitution $w = \sqrt{y}$, $dw = \frac{1}{2\sqrt{y}}\,dy$.

$$\int \frac{e^{\sqrt{y}}}{\sqrt{y}}\,dy = 2\int e^w\,dw = 2e^w + C = 2e^{\sqrt{y}} + C.$$

Check: $\frac{d}{dy}(2e^{\sqrt{y}} + C) = 2e^{\sqrt{y}}\cdot\frac{1}{2\sqrt{y}} = \frac{e^{\sqrt{y}}}{\sqrt{y}}.$

37. We use the substitution $w = 1 + 3t^2$, $dw = 6t\,dt$.

$$\int \frac{t}{1+3t^2}\,dt = \int \frac{1}{w}\left(\frac{1}{6}\,dw\right) = \frac{1}{6}\ln|w| + C = \frac{1}{6}\ln(1+3t^2) + C.$$

(We can drop the absolute value signs since $1 + 3t^2 > 0$ for all t).

Check: $\frac{d}{dt}\left[\frac{1}{6}\ln(1+3t^2) + C\right] = \frac{1}{6}\frac{1}{1+3t^2}(6t) = \frac{t}{1+3t^2}.$

Solutions for Section 7.3

1. Since $F'(x) = 5$, we use $F(x) = 5x$. By the Fundamental Theorem, we have

$$\int_1^3 5dx = 5x\bigg|_1^3 = 5(3) - 5(1) = 15 - 5 = 10.$$

5. Since $F'(x) = \frac{1}{x^2} = x^{-2}$, we use $F(x) = \frac{x^{-1}}{-1} = -\frac{1}{x}$. By the Fundamental Theorem, we have

$$\int_1^2 \frac{1}{x^2}dx = \left(-\frac{1}{x}\right)\bigg|_1^2 = -\frac{1}{2} - \left(-\frac{1}{1}\right) = -\frac{1}{2} + 1 = \frac{1}{2}.$$

9. If $F'(x) = 6x^2$, then $F(x) = 2x^3$. By the Fundamental Theorem, we have

$$\int_1^3 6x^2\,dx = 2x^3\bigg|_1^3 = 2(27) - 2(1) = 54 - 2 = 52.$$

13. If $f(x) = 1/x$, then $F(x) = \ln|x|$ (since $\frac{d}{dx}\ln|x| = \frac{1}{x}$). By the Fundamental Theorem, we have

$$\int_1^2 \frac{1}{x}\,dx = \ln|x|\bigg|_1^2 = \ln 2 - \ln 1 = \ln 2.$$

17. If $f(t) = e^{-0.2t}$, then $F(t) = -5e^{-0.2t}$. (This can be verified by observing that $\frac{d}{dt}(-5e^{-0.2t}) = e^{-0.2t}$.) By the Fundamental Theorem, we have

$$\int_0^1 e^{-0.2t}\,dt = (-5e^{-0.2t})\bigg|_0^1 = -5(e^{-0.2}) - (-5)(1) = 5 - 5e^{-0.2} \approx 0.906.$$

21. **(a)** We substitute $w = 1 + x^2$, $dw = 2x\,dx$.

$$\int_{x=0}^{x=1} \frac{x}{1+x^2}\,dx = \frac{1}{2}\int_{w=1}^{w=2} \frac{1}{w}\,dw = \frac{1}{2}\ln|w|\bigg|_1^2 = \frac{1}{2}\ln 2.$$

(b) We substitute $w = \cos x$, $dw = -\sin x\,dx$.

$$\int_{x=0}^{x=\frac{\pi}{4}} \frac{\sin x}{\cos x}\,dx = -\int_{w=1}^{w=\sqrt{2}/2} \frac{1}{w}\,dw$$

$$= -\ln|w|\bigg|_1^{\sqrt{2}/2} = -\ln\frac{\sqrt{2}}{2} = \frac{1}{2}\ln 2.$$

25. We have

$$\text{Area} = \int_1^4 x^2\,dx = \frac{x^3}{3}\bigg|_1^4 = \frac{4^3}{3} - \frac{1^3}{3} = \frac{64-1}{3} = 21.$$

29. To find the area under the graph of $f(x) = xe^{x^2}$, we need to evaluate the definite integral

$$\int_0^2 xe^{x^2}\,dx.$$

This is done in Example 4, Section 7.2, using the substitution $w = x^2$, the result being

$$\int_0^2 xe^{x^2}\,dx = \frac{1}{2}(e^4 - 1).$$

33. **(a)** At time $t = 0$, the rate of oil leakage $= r(0) = 50$ thousand liters/minute.
At $t = 60$, rate $= r(60) = 15.06$ thousand liters/minute.

(b) To find the amount of oil leaked during the first hour, we integrate the rate from $t = 0$ to $t = 60$:

$$\begin{aligned}\text{Oil leaked} &= \int_0^{60} 50e^{-0.02t}\,dt = \left(-\frac{50}{0.02}e^{-0.02t}\right)\Bigg|_0^{60} \\ &= -2500e^{-1.2} + 2500e^{0} = 1747 \text{ thousand liters.}\end{aligned}$$

37. **(a)** A calculator or computer gives

$$\int_1^{100} \frac{1}{\sqrt{x}}\,dx = 18 \qquad \int_1^{1000} \frac{1}{\sqrt{x}}\,dx = 61.2 \qquad \int_1^{10000} \frac{1}{\sqrt{x}}\,dx = 198.$$

These values do not seem to be converging.

(b) An antiderivative of $F'(x) = \dfrac{1}{\sqrt{x}}$ is $F(x) = 2\sqrt{x}$ $\left(\text{since} \dfrac{d}{dx}(2\sqrt{x}) = \dfrac{1}{\sqrt{x}}\right)$. So, by the Fundamental Theorem, we have

$$\int_1^{b} \frac{1}{\sqrt{x}}\,dx = 2\sqrt{x}\Big|_1^{b} = 2\sqrt{b} - 2\sqrt{1} = 2\sqrt{b} - 2.$$

(c) The limit of $2\sqrt{b} - 2$ as $b \to \infty$ does not exist, as $\sqrt{b}$ grows without bound. Therefore

$$\lim_{b\to\infty} \int_1^{b} \frac{1}{\sqrt{x}}\,dx = \lim_{b\to\infty} (2\sqrt{b} - 2) \quad \text{does not exist.}$$

So the improper integral $\displaystyle\int_1^{\infty} \frac{1}{\sqrt{x}}\,dx$ does not converge.

Solutions for Section 7.4

1. Since $F(0) = 0$, $F(b) = \int_0^b f(t)\,dt$. For each b we determine $F(b)$ graphically as follows:
$F(0) = 0$
$F(1) = F(0) + \text{Area of } 1 \times 1 \text{ rectangle} = 0 + 1 = 1$
$F(2) = F(1) + \text{Area of triangle } (\frac{1}{2} \cdot 1 \cdot 1) = 1 + 0.5 = 1.5$
$F(3) = F(2) + \text{Negative of area of triangle} = 1.5 - 0.5 = 1$
$F(4) = F(3) + \text{Negative of area of rectangle} = 1 - 1 = 0$
$F(5) = F(4) + \text{Negative of area of rectangle} = 0 - 1 = -1$
$F(6) = F(5) + \text{Negative of area of triangle} = -1 - 0.5 = -1.5$
The graph of $F(t)$, for $0 \le t \le 6$, is shown in Figure 7.1.

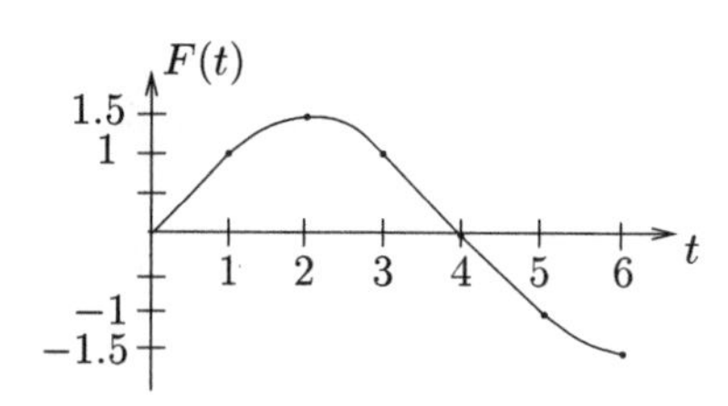

Figure 7.1

5.

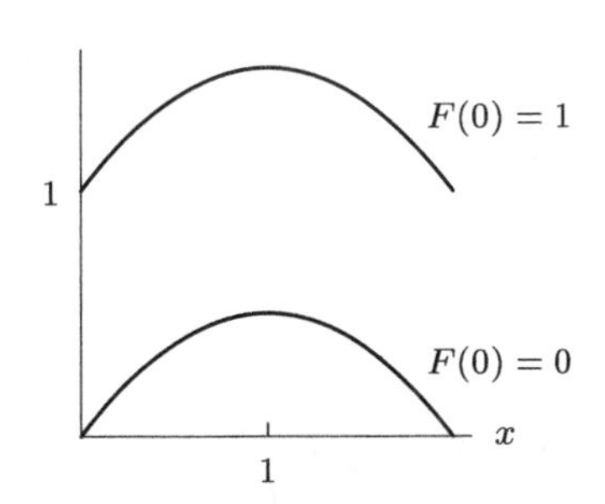

9. **(a)** The function $f(x)$ is increasing when $f'(x)$ is positive, so $f(x)$ is increasing for $-1 < x < 3$ or $x > 3$.
The function $f(x)$ is decreasing when $f'(x)$ is negative, so $f(x)$ is decreasing for $x < -1$.
Since $f(x)$ is decreasing to the left of $x = -1$ and increasing to the right of $x = -1$, the function has a local minimum at $x = -1$. Since $f(x)$ is increasing on both sides of $x = 3$, it has neither a local maximum nor a local minimum at that point.

(b) See Figure 7.2.

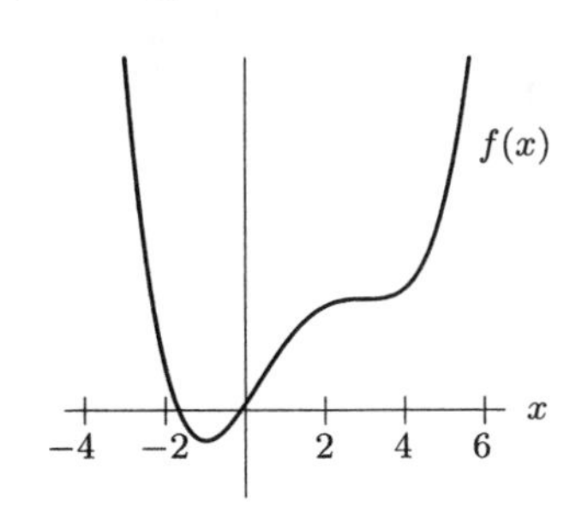

Figure 7.2

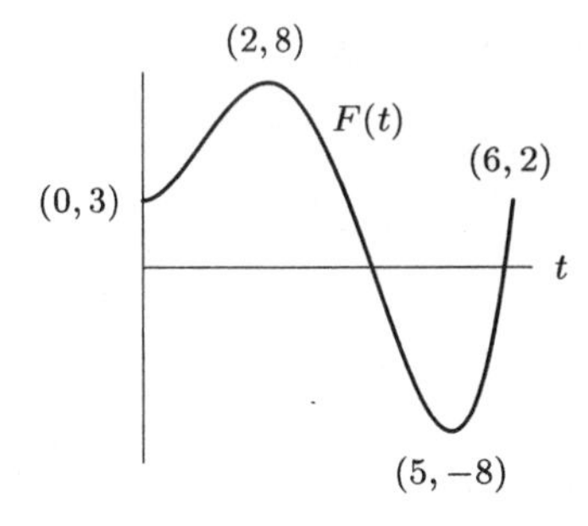

Figure 7.3

13. The areas given enable us to calculate the changes in the function F as we move along the t-axis. Areas above the axis count positively and areas below the axis count negatively. We know that $F(0) = 3$, so

$$F(2) - F(0) = \int_0^2 F'(t)\,dt = \begin{array}{c}\text{Area under } F' \\ 0 \le t \le 2\end{array} = 5$$

Thus,

$$F(2) = F(0) + 5 = 3 + 5 = 8.$$

Similarly,

$$F(5) - F(2) = \int_2^5 F'(t)\,dt = -16$$
$$F(5) = F(2) - 16 = 8 - 16 = -8$$

and

$$F(6) = F(5) + \int_5^6 F'(t)\,dt = -8 + 10 = 2.$$

A graph is shown in Figure 7.3.

17. The critical points are at $(0, 5)$, $(2, 21)$, $(4, 13)$, and $(5, 15)$. A graph is given below.

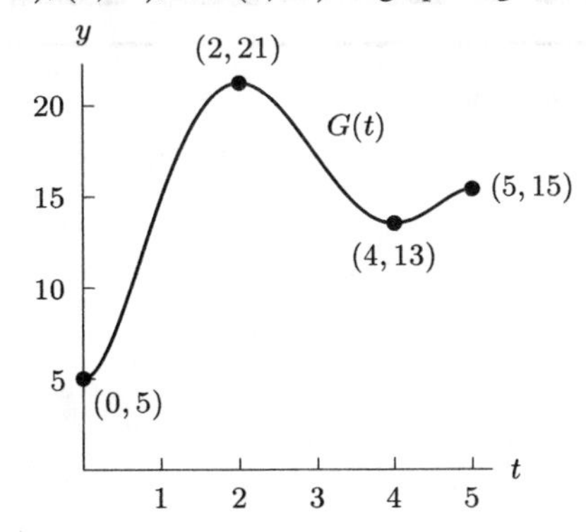

21. First rewrite each of the quantities in terms of f', since we have the graph of f'. If A_1 and A_2 are the positive areas shown in Figure 7.4:

$$f(3) - f(2) = \int_2^3 f'(t)\,dt = -A_1$$

$$f(4) - f(3) = \int_3^4 f'(t)\,dt = -A_2$$

$$\frac{f(4) - f(2)}{2} = \frac{1}{2}\int_2^4 f'(t)\,dt = -\frac{A_1 + A_2}{2}$$

Since Area $A_1 >$ Area A_2,

$$A_2 < \frac{A_1 + A_2}{2} < A_1$$

so

$$-A_1 < -\frac{A_1 + A_2}{2} < -A_2$$

and therefore

$$f(3) - f(2) < \frac{f(4) - f(2)}{2} < f(4) - f(3).$$

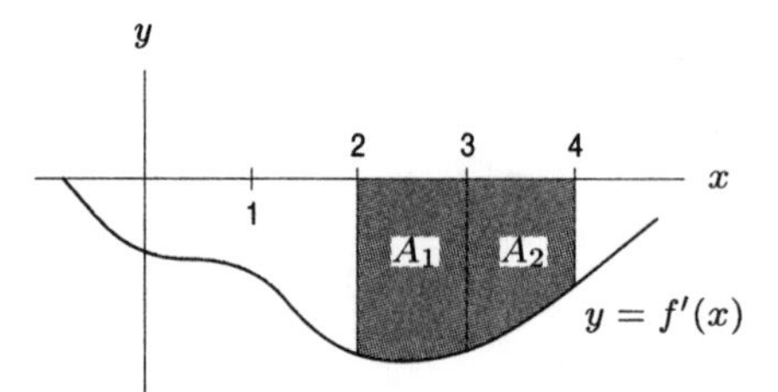

Figure 7.4

25.

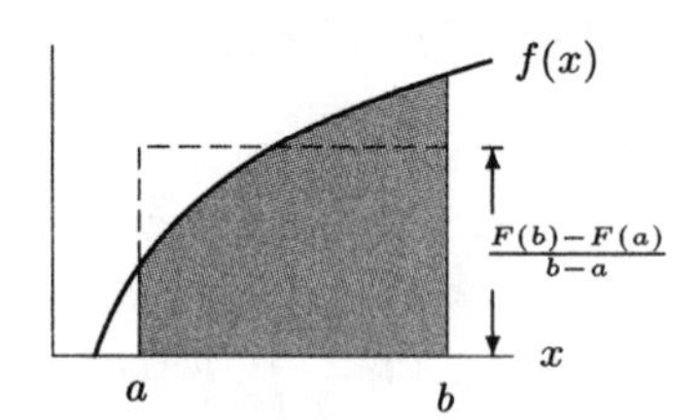

Note that we are using the interpretation of the definite integral as the length of the interval times the average value of the function on that interval, which we developed in Section 6.1.

Solutions for Chapter 7 Review

1. $10x + 8(\frac{x^4}{4}) = 10x + 2x^4$.

5. $2x^3 - 4x^2 + 3x$.

9. $P(r) = \pi r^2 + C$

13. $G(t) = 5t + \sin t + C$

17. $3x^3 + C$.

21. $e^x + 5x + C$

25. $2x^4 + \ln|x| + C$.

29. If $f(t) = 3t^2 + 4t + 3$, then $F(t) = t^3 + 2t^2 + 3t$. By the Fundamental Theorem, we have

$$\int_0^2 (3t^2 + 4t + 3)\,dt = (t^3 + 2t^2 + 3t)\Big|_0^2 = 2^3 + 2(2^2) + 3(2) - 0 = 22.$$

33. Since $F'(t) = 1/t^2 = t^{-2}$, we take $F(t) = \dfrac{t^{-1}}{-1} = -1/t$. Then

$$\begin{aligned}\int_1^2 \frac{1}{t^2}\,dt &= F(2) - F(1)\\ &= -\frac{1}{2} - \left(-\frac{1}{1}\right)\\ &= \frac{1}{2}.\end{aligned}$$

37. $f(x) = \sin x$, so $F(x) = -\cos x + C$. $F(0) = 0$ implies that $-\cos 0 + C = 0$, so $C = 1$. Thus $F(x) = -\cos x + 1$ is the only possibility.

41. We use the substitution $w = 3x + 1$, $dw = 3dx$.

$$\int \frac{1}{(3x+1)^2}dx = \frac{1}{3}\int \frac{1}{w^2}dw = \frac{1}{3}\int w^{-2}dw = \frac{1}{3}\frac{w^{-1}}{-1} + C = -\frac{1}{3(3x+1)} + C.$$

45. We use the substitution $w = 3x^2 + 4$, $dw = 6xdx$.

$$\int x\sqrt{3x^2+4}dx = \frac{1}{6}\int w^{1/2}dw = \frac{1}{6}\frac{w^{3/2}}{3/2} + C = \frac{1}{9}(3x^2+4)^{3/2} + C.$$

49. **(a)** In 1990, we have $P = 5.3e^{0.014(0)} = 5.3$ billion people.
In 2000, we have $P = 5.3e^{0.014(10)} = 6.1$ billion people.

(b) We have

$$\begin{aligned}\text{Average population} &= \frac{1}{10-0}\int_0^{10} 5.3e^{0.014t}dt = \frac{1}{10}\cdot\frac{5.3}{0.014}e^{0.014t}\Big|_0^{10}\\ &= \frac{1}{10}\left(\frac{5.3}{0.014}(e^{0.14} - e^0)\right) = 5.7.\end{aligned}$$

The average population of the world during the 1990s was 5.7 billion people.

53. **(a)** Critical points of $F(x)$ are $x = -1$, $x = 1$ and $x = 3$.

(b) $F(x)$ has a local minimum at $x = -1$, a local maximum at $x = 1$, and a local minimum at $x = 3$.

(c)

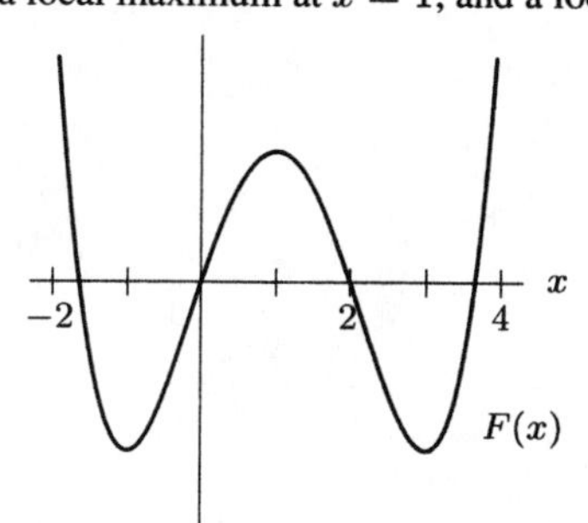

57. **(a)** Using a calculator or computer, we get

$$\int_0^3 e^{-2t}\,dt = 0.4988 \qquad \int_0^5 e^{-2t}\,dt = 0.49998$$

$$\int_1^7 e^{-2t}\,dt = 0.4999996 \qquad \int_0^{10} e^{-2t}\,dt = 0.499999999.$$

The values of these integrals are getting closer to 0.5. A reasonable guess is that the improper integral converges to 0.5.

(b) Since $-\frac{1}{2}e^{-2t}$ is an antiderivative of e^{-2t}, we have

$$\int_0^b e^{-2t}\,dt = -\frac{1}{2}e^{-2t}\Big|_0^b = -\frac{1}{2}e^{-2b} - \left(-\frac{1}{2}e^0\right) = -\frac{1}{2}e^{-2b} + \frac{1}{2}.$$

(c) Since $e^{-2b} = 1/e^{2b}$, we have

$$e^{2b} \to \infty \quad \text{as} \quad b \to \infty, \quad \text{so} \quad e^{-2b} = \frac{1}{e^{2b}} \to 0.$$

Therefore,

$$\lim_{b\to\infty} \int_0^b e^{-2t}\,dt = \lim_{b\to\infty} \left(-\frac{1}{2}e^{-2b} + \frac{1}{2}\right) = 0 + \frac{1}{2} = \frac{1}{2}.$$

So the improper integral converges to $1/2 = 0.5$:

$$\int_0^\infty e^{-2t}\,dt = \frac{1}{2}.$$

Solutions to Practice Problems on Integration

1. $\displaystyle\int (t^3 + 6t^2)\,dt = \frac{t^4}{4} + 6\cdot\frac{t^3}{3} + C = \frac{t^4}{4} + 2t^3 + C$

5. $\displaystyle\int 3w^{1/2}\,dw = 3\cdot\frac{w^{3/2}}{3/2} + C = 2w^{3/2} + C$

9. $\displaystyle\int (w^4 - 12w^3 + 6w^2 - 10)\,dw = \frac{w^5}{5} - 12\cdot\frac{w^4}{4} + 6\cdot\frac{w^3}{3} - 10\cdot w + C$

$$= \frac{w^5}{5} - 3w^4 + 2w^3 - 10w + C$$

13. $\displaystyle\int \left(\frac{4}{x} + 5x^{-2}\right)\,dx = 4\ln|x| + \frac{5x^{-1}}{-1} + C = 4\ln|x| - \frac{5}{x} + C$

17. $\displaystyle\int (5\sin x + 3\cos x)\,dx = -5\cos x + 3\sin x + C$

21. $\displaystyle\int 15p^2q^4\,dp = 15\left(\frac{p^3}{3}\right)q^4 + C = 5p^3q^4 + C$

25. $\displaystyle\int 5e^{2q}\,dq = 5\cdot\frac{1}{2}e^{2q} + C = 2.5e^{2q} + C$

29. $\displaystyle\int (x^2 + 8 + e^x)\,dx = \frac{x^3}{3} + 8x + e^x + C$

33. $\displaystyle\int (Aq + B)dq = \frac{Aq^2}{2} + Bq + C$

37. $\displaystyle\int 12\cos(4x)dx = 3\sin(4x) + C$

41. We use the substitution $w = y + 2$, $dw = dy$:

$$\int \frac{1}{y+2}dy = \int \frac{1}{w}dw = \ln|w| + C = \ln|y+2| + C.$$

45. We use the substitution $w = 1 + \sin x$, $dw = \cos x dx$:

$$\int \frac{\cos x}{\sqrt{1+\sin x}}dx = \int w^{-1/2}dw = \frac{w^{1/2}}{1/2} + C = 2\sqrt{1+\sin x} + C.$$

CHAPTER EIGHT

Solutions for Section 8.1

1. Since $p(x)$ is a density function,

$$\text{Area under graph} = \frac{1}{2} \cdot 50c = 25c = 1,$$

so $c = 1/25 = 0.04$.

5. We use the fact that the area of a rectangle is Base × Height.

(a) The fraction less than 5 meters high is the area to the left of 5, so

$$\text{Fraction} = 5 \cdot 0.05 = 0.25.$$

(b) The fraction above 6 meters high is the area to the right of 6, so

$$\text{Fraction} = (20 - 6)0.05 = 0.7.$$

(c) The fraction between 2 and 5 meters high is the area between 2 and 5, so

$$\text{Fraction} = (5 - 2)0.05 = 0.15.$$

9. If x is the yield in kg, the density function is a horizontal line at $p(x) = 1/100$ for $0 \leq x \leq 100$. See Figure 8.1.

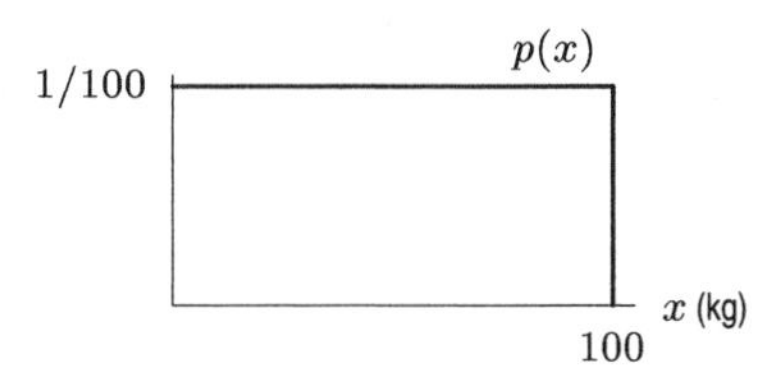

Figure 8.1

13. We can determine the fractions by estimating the area under the curve. Counting the squares for insects in the larval stage between 10 and 12 days we get 4.5 squares, with each square representing (2)·(3%) giving a total of 27% of the insects in the larval stage between 10 and 12 days.
Likewise we get 2 squares for the insects in the larval stage for less than 8 days, giving 12% of the insects in the larval stage for less than 8 days.
Likewise we get 7.5 squares for the insects in the larval stage for more than 12 days, giving 45% of the insects in the larval stage for more than 12 days.
Since the peak of the graph occurs between 12 and 13 days, the length of the larval stage is most likely to fall in this interval.

17. (a) The total area under the graph must be 1, so

$$\text{Area } = 5(0.01) + 5C = 1$$

So

$$5C = 1 - 0.05 = 0.95$$
$$C = 0.19$$

(b) The machine is more likely to break in its tenth year than first year. It is equally likely to break in its first year and second year.

(c) Since $p(t)$ is a density function,

$$\begin{aligned}
\text{Fraction of machines lasting up to 2 years} &= \text{Area from 0 to 2} = 2(0.01) = 0.02. \\
\text{Fraction of machines lasting between 5 and 7 years} &= \text{Area from 5 to 7} = 2(0.19) = 0.38. \\
\text{Fraction of machines lasting between 3 and 6 years} &= \text{Area from 3 to 6} \\
&= \text{Area from 3 to 5} + \text{Area from 5 to 6} \\
&= 2(0.01) + 1(0.19) = 0.21.
\end{aligned}$$

Solutions for Section 8.2

1.

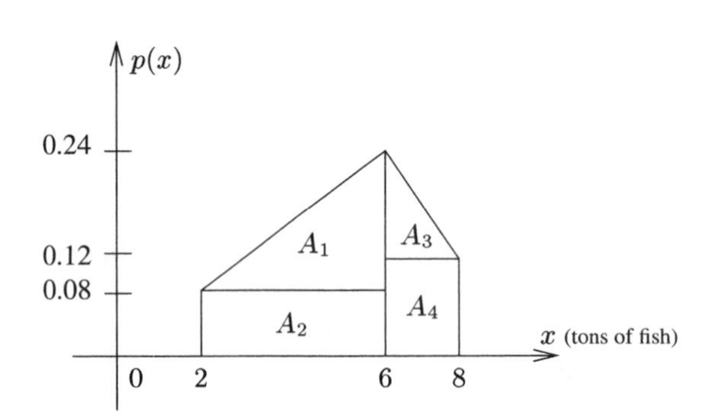

Splitting the figure into four pieces, we see that

$$\begin{aligned}
\text{Area under the curve} &= A_1 + A_2 + A_3 + A_4 \\
&= \frac{1}{2}(0.16)4 + 4(0.08) + \frac{1}{2}(0.12)2 + 2(0.12) \\
&= 1.
\end{aligned}$$

We expect the area to be 1, since $\int_{-\infty}^{\infty} p(x)\,dx = 1$ for any probability density function, and $p(x)$ is 0 except when $2 \le x \le 8$.

5.

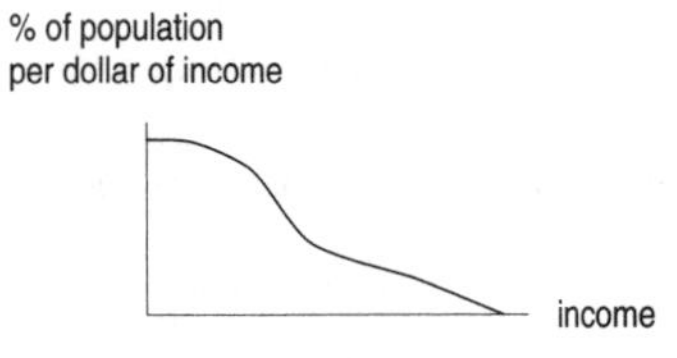

Figure 8.2: Density function

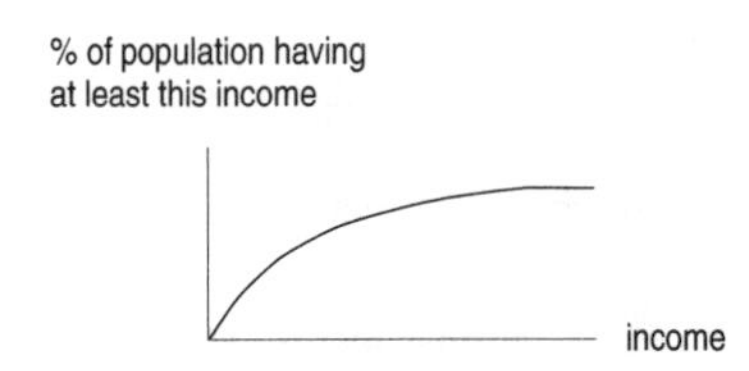

Figure 8.3: Cumulative distribution function

9. (a) The first item is sold at the point at which the graph is first greater than zero. Thus the first item is sold at $t = 30$ or January 30. The last item is sold at the t value at which the function is first equal to 100%. Thus the last item is sold at $t = 240$ or August 28, unless its a leap year.

(b) Looking at the graph at $t = 121$ we see that roughly 65% of the inventory has been sold by May 1.

(c) The percent of the inventory sold during May and June is the difference between the percent of the inventory sold on the last day of June and the percent of the inventory sold on the first day of May. Thus, the percent of the inventory sold during May and June is roughly 25%.

(d) The percent of the inventory left after half a year is

$$100 - (\text{percent inventory sold after half year}).$$

Thus, roughly 10% of the inventory is left after half a year.

(e) The items probably went on sale on day 100 and were on sale until day 120. Roughly from April 10 until April 30.

13. (a) The fraction of students passing is given by the area under the curve from 2 to 4 divided by the total area under the curve. This appears to be about $\frac{2}{3}$.

(b) The fraction with honor grades corresponds to the area under the curve from 3 to 4 divided by the total area. This is about $\frac{1}{3}$.

(c) The peak around 2 probably exists because many students work to get just a passing grade.

(d)

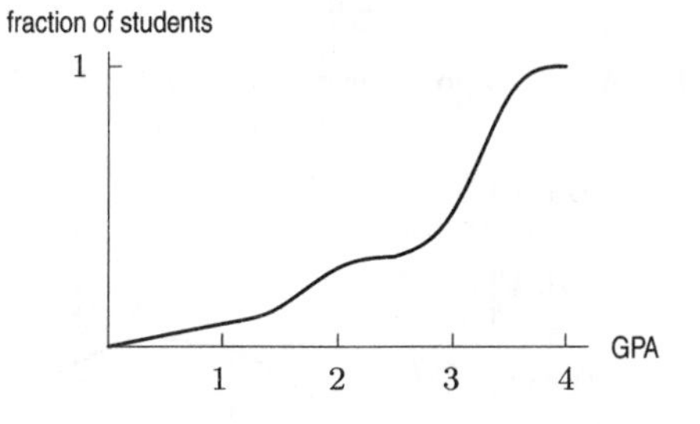

17. (a) The probability that a banana lasts between 1 and 2 weeks is given by

$$\int_1^2 p(t)dt = 0.25$$

Thus there is a 25% probability that the banana will last between one and two weeks.

(b) The formula given for $p(t)$ is valid for up to four weeks; for $t > 4$ we have $p(t) = 0$. So a banana lasting more than 3 weeks must last between 3 and 4 weeks. Thus the probability is

$$\int_3^4 p(t)dt = 0.325$$

32.5% of the bananas last more than 3 weeks.

(c) Since $p(t) = 0$ for $t > 4$, the probability that a banana lasts more than 4 weeks is 0.

Solutions for Section 8.3

1. The median daily catch is the amount of fish such that half the time a boat will bring back more fish and half the time a boat will bring back less fish. Thus the area under the curve and to the left of the median must be 0.5. There are 25 squares under the curve so the median occurs at 12.5 squares of area. Now

$$\int_2^5 p(x)dx = 10.5 \text{ squares}$$

and

$$\int_5^6 p(x)dx = 5.5 \text{ squares},$$

so the median occurs at a little over 5 tons. We must find the value a for which

$$\int_5^a p(t)dt = 2 \text{ squares},$$

and we note that this occurs at about $a = 0.35$. Hence

$$\begin{aligned}\int_2^{5.35} p(t)\,dt &\approx 12.5 \text{ squares} \\ &\approx 0.5.\end{aligned}$$

The median is about 5.35 tons.

5. We know that the mean is given by

$$\int_{-\infty}^{\infty} tp(t)dt.$$

Thus we get

$$\begin{aligned}\text{Mean } &= \int_0^4 tp(t)dt \\ &= \int_0^4 (-0.0375t^3 + 0.225t^2)dt \\ &\approx 2.4\end{aligned}$$

Thus the mean is 2.4 weeks. Figure 8.4 supports the conclusion that $t = 2.4$ is in fact the mean.

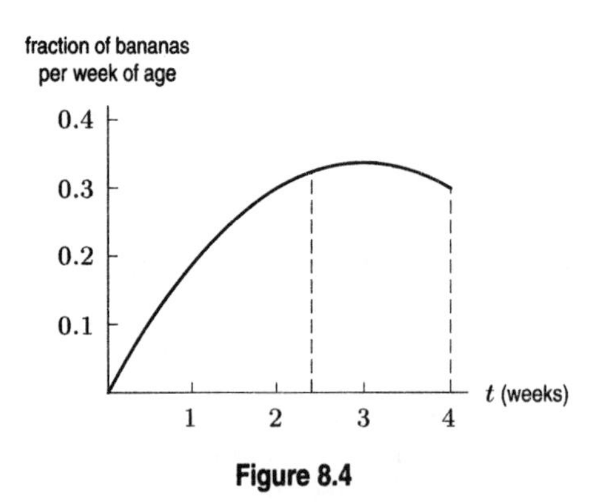

Figure 8.4

9. **(a)** P is the cumulative distribution function, so the percentage of the population that made between \$20,000 and \$50,000 is

$$P(50) - P(20) = 99\% - 75\% = 24\%.$$

Therefore $\frac{6}{25}$ of the population made between \$20,000 and \$50,000.

(b) The median income is the income such that half the people made less than this amount. Looking at the chart, we see that $P(12.6) = 50\%$, so the median must be \$12,600.

(c) The cumulative distribution function looks something like Figure 8.5. The density function is the derivative of the cumulative distribution. Qualitatively it looks like Figure 8.6.

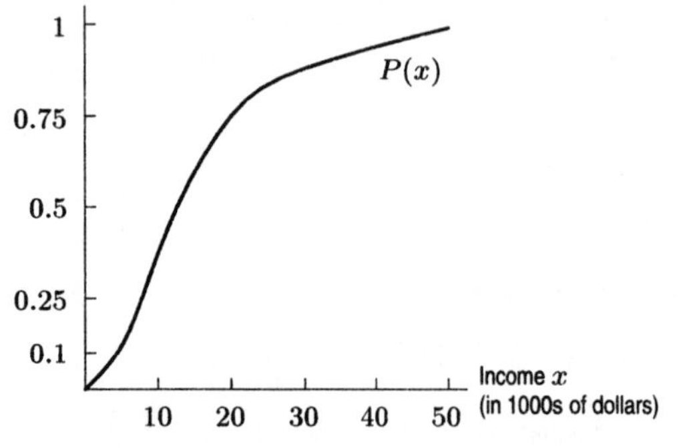

Figure 8.5: Cumulative distribution

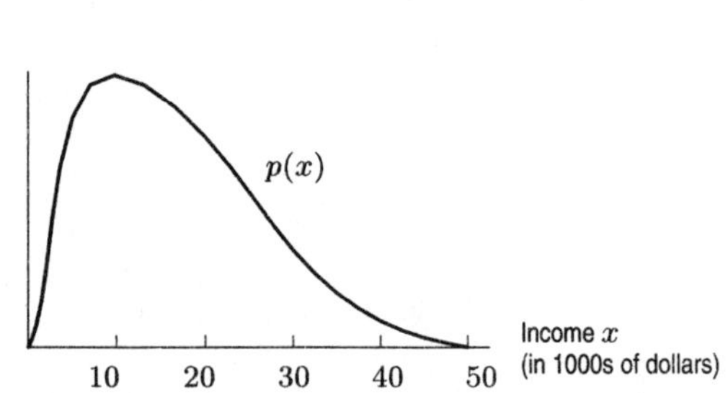

Figure 8.6: Density function

The density function has a maximum at about \$8000. This means that more people have incomes around \$8000 than around any other amount. On the density function, this is the highest point. On the cumulative distribution, this is the point of steepest slope (because $P' = p$), which is also the point of inflection.

Solutions for Chapter 8 Review

1. Suppose x is the age of death; a possible density function is graphed in Figure 8.7.

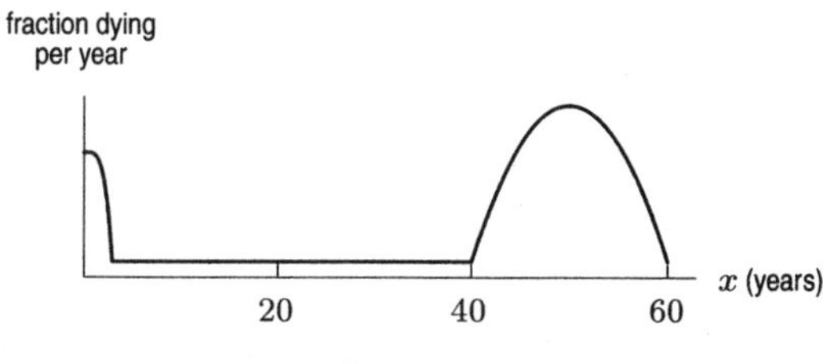

Figure 8.7

5. Since $p(x)$ is a density function, the area under the graph of $p(x)$ is 1, so

$$\text{Area} = \frac{1}{2}\text{Base} \cdot \text{Height} = \frac{1}{2} \cdot 10 \cdot a = 5a = 1$$
$$a = \frac{1}{5}.$$

9. **(a)** See Figure 8.8. This is a cumulative distribution function.

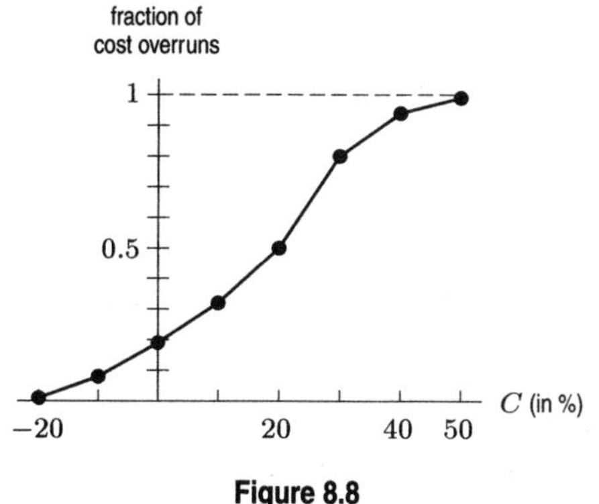

Figure 8.8

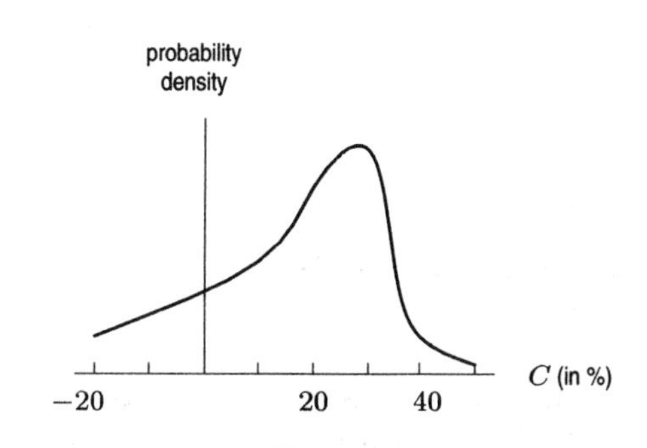

Figure 8.9

(b) The density function is the derivative of the cumulative distribution function. See Figure 8.9.

(c) Let's call the cumulative distribution function $F(C)$. The probability that there will be a cost overrun of more than 50% is $1 - F(50) = 0.01$, a 1% chance. The probability that it will be between 20% and 50% is $F(50) - F(20) = 0.99 - 0.50 = 0.49$, or 49%. The most likely amount of cost overrun occurs when the slope of the tangent line to the cumulative distribution function is a maximum. This occurs at the inflection point of the cumulative distribution graph, at about $C = 28\%$.

13. **(a)** Since $d(e^{-ct})/dt = ce^{-ct}$, we have

$$c\int_0^6 e^{-ct}dt = -e^{-ct}\Big|_0^6 = 1 - e^{-6c} = 0.1,$$

so

$$c = -\frac{1}{6}\ln 0.9 \approx 0.0176.$$

(b) Similarly, with $c = 0.0176$, we have

$$c\int_6^{12} e^{-ct}dt = -e^{-ct}\Big|_6^{12}$$
$$= e^{-6c} - e^{-12c} = 0.9 - 0.81 = 0.09,$$

so the probability is 9%.

17. True. The interval from $x = 9.98$ to $x = 10.04$ has length 0.06. Assuming that the value of $p(x)$ is near $1/2$ for $9.98 < x < 10.04$, the fraction of the population in that interval is $\int_{9.98}^{10.04} p(x)dx \approx (1/2)(0.06) = 0.03$.

CHAPTER NINE

Solutions for Section 9.1

1. Beef consumption by households making \$20,000/year is given by Row 1 of Table 9.2 on page 324 of the text.

Table 9.1

p	3.00	3.50	4.00	4.50
$f(20,p)$	2.65	2.59	2.51	2.43

For households making \$20,000/year, beef consumption decreases as price goes up.
Beef consumption by households making \$100, 000/year is given by Row 5 of Table 9.2.

Table 9.2

p	3.00	3.50	4.00	4.50
$f(100,p)$	5.79	5.77	5.60	5.53

For households making \$100,000/year, beef consumption also decreases as price goes up.
Beef consumption by households when the price of beef is \$3.00/lb is given by Column 1 of Table 9.2.

Table 9.3

I	20	40	60	80	100
$f(I,3.00)$	2.65	4.14	5.11	5.35	5.79

When the price of beef is \$3.00/lb, beef consumption increases as income increases.
Beef consumption by households when the price of beef is \$4.00/lb is given by Column 3 of Table 9.2.

Table 9.4

I	20	40	60	80	100
$f(I,4.00)$	2.51	3.94	4.97	5.19	5.60

When the price of beef is \$4.00/lb, beef consumption increases as income increases.

5. In the answer to Problem 4 we saw that

$$P = 0.052\frac{M}{I},$$

and in the answer to Problem 3 we saw that

$$M = pf(I,p).$$

Putting the expression for M into the expression for P, gives:

$$P = 0.052\frac{pf(I,p)}{I}.$$

9. To see whether f is an increasing or decreasing function of x, we need to see how f varies as we increase x and hold y fixed. We note that each column of the table corresponds to a fixed value of y. Scanning down the $y = 2$ column, we can see that as x increases, the value of the function decreases from 114 when $x = 0$ down to 93 when $x = 80$. Thus, f may be decreasing. In order for f to actually be decreasing however, we have to make sure that f decreases for *every* column. In this case, we see that f indeed does decrease for every column. Thus, f is a decreasing function of x. Similarly, to see whether f is a decreasing function of y we need to look at the rows of the table. As we can see, f increases for every row as we increase y. Thus, f is an increasing function of y.

13. **(a)** According to Table 9.4 of the problem, it feels like $-31°$F.
(b) A wind of 10 mph, according to Table 9.4.
(c) About 5.5 mph. Since at a temperature of 25°F, when the wind increases from 5 mph to 10 mph, the temperature adjusted for wind-chill decreases from 21°F to 10°F, we can say that a 5 mph increase in wind speed causes an 11°F decrease in the temperature adjusted for wind-chill. Thus, each 0.5 mph increase in wind speed brings *about* a 1°F drop in the temperature adjusted for wind-chill.
(d) With a wind of 15 mph, approximately 23.5°F would feel like 0°F. With a 15 mph wind speed, when air temperature drops five degrees from 25°F to 20°F, the temperature adjusted for wind-chill drops 7 degrees from 2°F to $-5°$F. We can say that for every 1°F decrease in temperature there is *about* a 1.4°F ($= 7/5$) drop in the temperature you feel.

17.

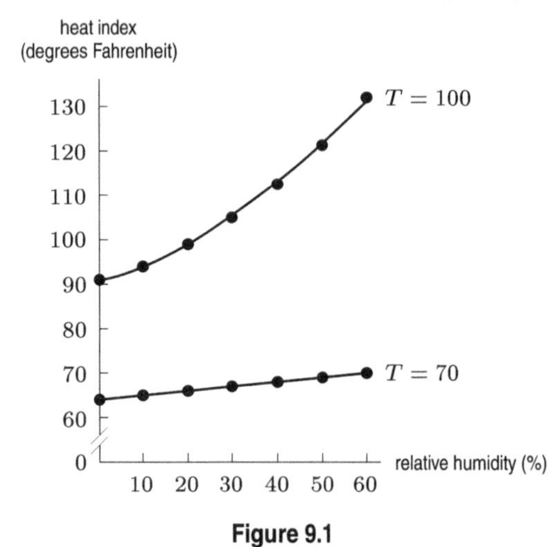

Figure 9.1

Both graphs are increasing because at any fixed temperature the air feels hotter as the humidity increases. The fact that the graph for $T = 100$ increases more rapidly with humidity than the graph for $T = 70$ tells us that when it is hot (100°F), high humidity has more effect on how we feel than at lower temperatures (70°F).

21. It stands to reason that the demand for tea, D, will have a similar formula to the demand for coffee. The demand D will be a decreasing function of the price of tea, t, and an increasing function of the price of coffee c. A possible formula for the demand for tea is

$$D = 100\frac{c}{t}.$$

Solutions for Section 9.2

1. **(a)** 80-90°F
(b) 60-72°F
(c) 60-100°F

5. We can see that as we move horizontally to the right, we are increasing x but not changing y. As we take such a path at $y = 2$, we cross decreasing contour lines, starting at the contour line 6 at $x = 1$ to the contour line 1 at around $x = 5.7$. This trend holds true for all of horizontal paths. Thus, z is a decreasing function of x. Similarly, as we move up along a vertical line, we cross increasing contour lines and thus z is an increasing function of y.

9. If we let x represent the high temperature and y represent the low temperature, then the contours should be highest for large x values and small y values. The contour values are greatest in the lower right corner of Quadrant I. One possible contour diagram is shown in Figure 9.2. Other answers are also possible.

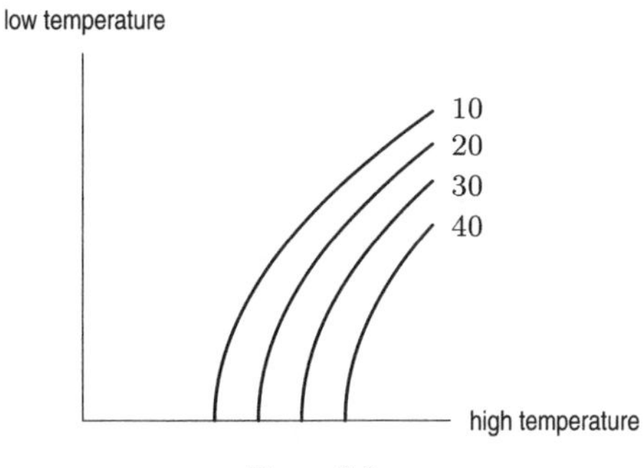

Figure 9.2

13. **(a)** The point representing 13% and \$6000 on the graph lies between the 120 and 140 contours. We estimate the monthly payment to be about \$137.

(b) Since the interest rate has dropped, we will be able to borrow more money and still make a monthly payment of \$137. To find out how much we can afford to borrow, we find where the interest rate of 11% intersects the \$137 contour and read off the loan amount to which these values correspond. Since the \$137 contour is not shown, we estimate its position from the \$120 and \$140 contours. We find that we can borrow an amount of money that is more than \$6000 but less than \$6500. So we can borrow about \$250 more without increasing the monthly payment.

(c) The entries in the table will be the amount of loan at which each interest rate intersects the 137 contour. Using the \$137 contour from (b) we make table 9.5.

Table 9.5 *Amount borrowed at a monthly payment of \$137.*

Interest Rate (%)	0	1	2	3	4	5	6	7
Loan Amount (\$)	8200	8000	7800	7600	7400	7200	7000	6800
Interest rate (%)	8	9	10	11	12	13	14	15
Loan Amount (\$)	6650	6500	6350	6250	6100	6000	5900	5800

17. The contour where $f(x, y) = 3x + 3y = c$ or $y = -x + c/3$ is the graph of the straight line of slope -1 as shown in Figure 9.3. Note that we have plotted the contours for $c = -9, -6, -3, 0, 3, 6, 9$. The contours are evenly spaced.

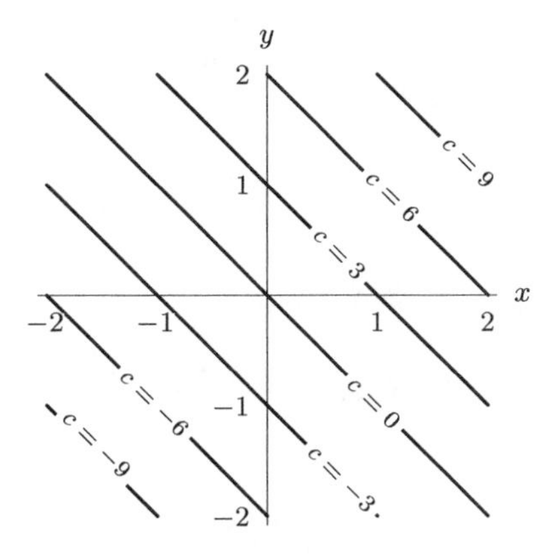

Figure 9.3

21. One possible answer follows.

(a) If the middle line is a highway in the city, there may be many people living around it since it provides mass transit nearby that nearly everyone needs and uses. Thus, the population density would be greatest near the highway, and less dense farther from the highway, as is shown in diagram (I). In an alternative solution, highways may be considered inherently very noisy and dirty. In this scenario, the people in a certain community may purposely live away from the highway. Then the population density would follow the pattern in diagram (III), or in an extreme case, even diagram (II).

(b) If the middle line is a sewage canal, there will be no one living within it, and very few people living close to it. This is represented in diagram (II).

(c) If the middle line is a railroad line in the city, then one scenario is that very few people would enjoy living nearby a railroad yard, with all of the noise and difficulty in crossing it. Then the population density would be smallest near the middle line, and greatest farther from the railroad, as in diagram (III). In an alternative solution, a community might well depend upon the railroad for transit, news or supplies, in which case there would be a denser population nearer the railroad than farther from it, as in diagram (I).

25. (a) The contour lines are much closer together on path A, so path A is steeper.

(b) If you are on path A and turn around to look at the countryside, you find hills to your left and right, obscuring the view. But the ground falls away on either side of path B, so you are likely to get a much better view of the countryside from path B.

(c) There is more likely to be a stream alongside path A, because water follows the direction of steepest descent.

29. The point $x = 10$, $t = 5$ is between the contours $H = 70$ and $H = 75$, a little closer to the former. Therefore, we estimate $H(10,5) \approx 72$, i.e., it is about 72°F. Five minutes later we are at the point $x = 10$, $t = 10$, which is just above the contour $H = 75$, so we estimate that it has warmed up to 76°F by then.

Solutions for Section 9.3

1. (a) We expect the demand for coffee to decrease as the price of coffee increases (assuming the price of tea is fixed.) Thus we expect f_c to be negative. We expect people to switch to coffee as the price of tea increases (assuming the price of coffee is fixed), so that the demand for coffee will increase. We expect f_t to be positive.

(b) The statement $f(3,2) = 780$ tells us that if coffee costs \$3 per pound and tea costs \$2 per pound, we can expect 780 pounds of coffee to sell each week. The statement $f_c(3,2) = -60$ tells us that, if the price of coffee then goes up \$1 and the price of tea stays the same, the demand for coffee will go down by about 60 pounds. The statement $20 = f_t(3,2)$ tells us that if the price of tea goes up \$1 and the price of coffee stays the same, the demand for coffee will go up by about 20 pounds.

5. (a) We expect f_p to be negative because if the price of the product increases, the sales usually decrease.

(b) If the price of the product is \$8 per unit and if \$12000 has been spent on advertising, sales increase by approximately 150 units if an additional \$1000 is spent on advertising.

9. (a) If we move horizontally by increasing x along any of the rows of this table, we see that the value of f decreases when x increases, so $f_x < 0$. Similarly, if we move vertically by increasing y along any of the columns of this table, we see that the value of f increases, so $f_y > 0$.

(b) We know that the formula for f_x at a point (x,y) is

$$f_x(x,y) \approx \frac{f(x+\Delta x, y) - f(x,y)}{\Delta x}.$$

Approximating with $\Delta x = 10$ at the point $(10,6)$ gives:

$$\begin{aligned} f_x(10,6) &\approx \frac{f(10+10,6) - f(10,6)}{10} \\ &= \frac{f(20,6) - f(10,6)}{10} \\ &= \frac{92-98}{10} = -0.6. \end{aligned}$$

Similarly for f_y, we approximate with $\Delta y = 2$.

$$\begin{aligned} f_y(10,6) &\approx \frac{f(10,6+2) - f(10,6)}{2} \\ &= \frac{f(10,8) - f(10,6)}{2} \\ &= \frac{105-98}{2} = 3.5. \end{aligned}$$

(c) We use $\Delta f \approx \Delta x f_x + \Delta y f_y$. To calculate $f(30,9)$ using the value for $f(30,8)$ shown in the table, we need to calculate the partial derivative $f_y(30,8)$. Using $\Delta y = -2$, we have

$$f_y(30,8) \approx \frac{f(30,8-2) - f(30,8)}{-2} = 3.$$

Then

$$\begin{aligned} f(30,9) &\approx f(30,8) + 0 \cdot f_x(30,8) + 1 \cdot f_y(30,8) \\ &\approx 94 + 0 + 3 \\ &= 97. \end{aligned}$$

To calculate $f(34,8)$ using the value for $f(30,8)$ shown in the table, we need to calculate the partial derivative $f_x(30,8)$. Using $\Delta x = -10$, we have

$$f_x(30,8) \approx \frac{f(30-10,8) - f(30,8)}{-10} = -0.5.$$

Then

$$\begin{aligned} f(34,8) &\approx f(30,8) + 4 \cdot f_x(30,8) + 0 \cdot f_y(30,8) \\ &\approx 94 - 2 + 0 \\ &= 92. \end{aligned}$$

13. (a) We know we can approximate f_p at a point (I,p) by $f_p(I,p) \approx \frac{f(I,p+\Delta p) - f(I,p)}{\Delta p}$. Approximating with $\Delta p = 0.5$ at the point $(80, 4.0)$ gives:

$$\begin{aligned} f_p(80,4.0) &\approx \frac{f(80,4.0+0.5) - f(80,4.0)}{0.5} \\ &= \frac{f(80,4.5) - f(80,4.0)}{0.5} \\ &= \frac{5.07 - 5.19}{0.5} = -0.24 \end{aligned}$$

When beef rises in price by a dollar per pound, the average household with an income of \$80,000 will buy 0.24 fewer pounds of beef per week.

(b) We estimate f_I similarly; we approximate with $\Delta I = 20$.

$$\begin{aligned} f_I(80,4.0) &\approx \frac{f(80+20,4.0) - f(80,4.0)}{20} \\ &= \frac{f(100,4.0) - f(80,4.0)}{20} \\ &= \frac{5.60 - 5.19}{20} = 0.0205 \end{aligned}$$

When a household's income rises by \$1000, it will buy 0.0205 more pounds of beef per week.

(c) We wish to find $f(110, 4.0)$. To do this, we employ the formula $\Delta f \approx \Delta I f_I + \Delta p f_p$. Using our results for $f_I(80,4.0)$ and $f_p(110,4.0)$ from above gives

$$\begin{aligned} f(110,4.0) &\approx f(80,4.0) + 30 \cdot f_I(80,4.0) + 0 \cdot f_p(80,4.0) \\ &\approx 5.60 + 0.615 + 0 \\ &= 6.215 \text{ lbs.} \end{aligned}$$

17. We can use the formula $\Delta f \approx \Delta x f_x + \Delta y f_y$. Applying this formula in order to estimate $f(105,21)$ from the known value of $f(100,20)$ gives

$$\begin{aligned} f(105,21) &\approx f(100,20) + (105-100)f_x(100,20) + (21-20)f_y(100,20) \\ &= 2750 + 5 \cdot 4 + 1 \cdot 7 \\ &= 2777. \end{aligned}$$

21. Since the average rate of change of the temperature adjusted for wind-chill is about -2.6 (drops by 2.6°F), with every 1 mph increase in wind speed from 5 mph to 10 mph, when the true temperature stays constant at 20°F, we know that

$$f_w(5, 20) \approx -2.6.$$

25. **(a)** The partial derivative $f_x = 350$ tells us that R increases by \$350 as x increases by 1. Thus, $f(201, 400) = f(200, 400) + 350 = 150{,}000 + 350 = 150{,}350$.

(b) The partial derivative $f_y = 200$ tells us that R increases by \$200 as y increases by 1. Since y is increasing by 5, we have $f(200, 405) = f(200, 400) + 5(200) = 150{,}000 + 1{,}000 = 151{,}000$.

(c) Here x is increasing by 3 and y is increasing by 6. We have $f(203, 406) = f(200, 400) + 3(350) + 6(200) = 150{,}000 + 1{,}050 + 1{,}200 = 152{,}250$.

In this problem, the partial derivatives gave exact results, but in general they only give an estimate of the changes in the function.

29. We know $z_x(1, 0)$ is the rate of change of z in the x-direction at $(1, 0)$.Therefore

$$z_x(1,0) \approx \frac{\Delta z}{\Delta x} \approx \frac{1}{0.5} = 2, \quad \text{so } z_x(1,0) \approx 2.$$

We know $z_x(0, 1)$ is the rate of change of z in the x-direction at the point $(0, 1)$. Since we move along the contour, the change in z

$$z_x(0,1) \approx \frac{\Delta z}{\Delta x} \approx \frac{0}{\Delta x} = 0.$$

We know $z_y(0, 1)$ is the rate of change of z in the y-direction at the point $(0, 1)$ so

$$z_y(0,1) \approx \frac{\Delta z}{\Delta y} \approx \frac{1}{0.1} = 10.$$

Solutions for Section 9.4

1.

$$f(1,2) = (1)^3 + 3(2)^2 = 13$$
$$f_x(x,y) = 3x^2 + 0 \Rightarrow f_x(1,2) = 3(1)^2 = 3$$
$$f_y(x,y) = 0 + 6y \Rightarrow f_y(1,2) = 6(2) = 12$$

5. $f_x(x, y) = 4x + 0 = 4x$
$f_y(x, y) = 0 + 6y = 6y$

9. $f_x(x, y) = 2 \cdot 100xy = 200xy$
$f_y(x, y) = 100x^2 \cdot 1 = 100x^2$

13. $\dfrac{\partial A}{\partial h} = \dfrac{1}{2}(a + b)$

17. We have

$$f_x(x,y) = 3x^2 + 6xy \quad \text{and} \quad f_y(x,y) = 3x^2 - 4y,$$

so

$$f_x(1,2) = 15 \quad \text{and} \quad f_y(1,2) = -5.$$

21.

$$\begin{aligned} f(500, 1000) &= 16 + 1.2(500) + 1.5(1000) + 0.2(500)(1000) \\ &= \$102{,}116 \end{aligned}$$

The cost of producing 500 units of item 1 and 1000 units of item 2 is $102,116.

$$f_{q_1}(q_1, q_2) = 0 + 1.2 + 0 + 0.2q_2$$

So $f_{q_1}(500, 1000) = 1.2 + 0.2(1000) = \$201.20$ per unit. When the company is producing at $q_1 = 500$, $q_2 = 1000$, the cost of producing one more unit of item 1 is $201.20.

$$f_{q_2}(q_1, q_2) = 0 + 0 + 1.5 + 0.2q_1$$

So $f_{q_2}(500, 1000) = 1.5 + 0.2(500) = \$101.50$ per unit. When the company is producing at $q_1 = 500$, $q_2 = 1000$, the cost of producing one more unit of item 2 is $101.50.

25. **(a)** $Q_K = 18.75K^{-0.25}L^{0.25}$, $Q_L = 6.25K^{0.75}L^{-0.75}$.

(b) When $K = 60$ and $L = 100$,

$$\begin{aligned} Q &= 25 \cdot 60^{0.75} \cdot 100^{0.25} = 1704.33 \\ Q_K &= 18.75 \cdot 60^{-0.25} 100^{0.25} = 21.3 \\ Q_L &= 6.25 \cdot 60^{0.75} 100^{-0.75} = 4.26 \end{aligned}$$

(c) Q is actual quantity being produced. Q_K is how much more could be produced if you increased K by one unit. Q_L is how much more could be produced if you increased L by 1.

29. $f_x = 2xy$ and $f_y = x^2$, so $f_{xx} = 2y$, $f_{xy} = 2x$, $f_{yy} = 0$ and $f_{yx} = 2x$.

33. $f_x = 2xy^2$ and $f_y = 2x^2y$, so $f_{xx} = 2y^2$, $f_{xy} = 4xy$, $f_{yy} = 2x^2$ and $f_{yx} = 4xy$.

37. $P_K = 2L^2$ and $P_L = 4KL$, so $P_{KK} = 0$, $P_{LL} = 4K$ and $P_{KL} = P_{LK} = 4L$.

41. Since $f_x(x, y) = 4x^3y^2 - 3y^4$, we could have

$$f(x, y) = x^4y^2 - 3xy^4.$$

In that case,

$$f_y(x, y) = \frac{\partial}{\partial y}(x^4y^2 - 3xy^4) = 2x^4y - 12xy^3$$

as expected. More generally, we could have $f(x, y) = x^4y^2 - 3xy^4 + C$, where C is any constant.

Solutions for Section 9.5

1. Mississippi lies entirely within a region designated as 80s so we expect both the maximum and minimum daily high temperatures within the state to be in the 80s. The southwestern-most corner of the state is close to a region designated as 90s, so we would expect the temperature here to be in the high 80s, say 87-88. The northern-most portion of the state is located near the center of the 80s region. We might expect the high temperature there to be between 83-87.

Alabama also lies completely within a region designated as 80s so both the high and low daily high temperatures within the state are in the 80s. The southeastern tip of the state is close to a 90s region so we would expect the temperature here to be about 88-89 degrees. The northern-most part of the state is near the center of the 80s region so the temperature there is 83-87 degrees.

Pennsylvania is also in the 80s region, but it is touched by the boundary line between the 80s and a 70s region. Thus we expect the low daily high temperature to occur there and be about 80 degrees. The state is also touched by a boundary line of a 90s region so the high will occur there and be 89-90 degrees.

New York is split by a boundary between an 80s and a 70s region, so the northern portion of the state is likely to be about 74-76 while the southern portion is likely to be in the low 80s, maybe 81-84 or so.

California contains many different zones. The northern coastal areas will probably have the daily high as low as 65-68, although without another contour on that side, it is difficult to judge how quickly the temperature is dropping off to the west. The tip of Southern California is in a 100s region, so there we expect the daily high to be 100-101.

Arizona will have a low daily high around 85-87 in the northwest corner and a high in the 100s, perhaps 102-107 in its southern regions.

Massachusetts will probably have a high daily high around 81-84 and a low daily high of 70.

5. At a critical point $f_x = -3y + 6 = 0$ and $f_y = 3y^2 - 3x = 0$, so $(4, 2)$ is the only critical point. Since $f_{xx} f_{yy} - f_{xy}^2 = -9 < 0$, the point $(4, 2)$ is neither a local maximum nor a local minimum.

9. To find the critical points, we solve $f_x = 0$ and $f_y = 0$ for x and y. Solving

$$f_x = 3x^2 - 6x = 0$$
$$f_y = 3y^2 - 3 = 0$$

shows that $x = 0$ or $x = 2$ and $y = -1$ or $y = 1$. There are four critical points: $(0, -1)$, $(0, 1)$, $(2, -1)$, and $(2, 1)$.

We have

$$D = (f_{xx})(f_{yy}) - (f_{xy})^2 = (6x - 6)(6y) - (0)^2 = (6x - 6)(6y).$$

At the point $(0, -1)$, we have $D > 0$ and $f_{xx} < 0$, so f has a local maximum.
At the point $(0, 1)$, we have $D < 0$, so f has neither a local maximum nor a local minimum.
At the point $(2, -1)$, we have $D < 0$, so f has neither a local maximum nor a local minimum.
At the point $(2, 1)$, we have $D > 0$ and $f_{xx} > 0$, so f has a local minimum.

13. We can identify local extreme points on contour plots because these points will be the centers of a series of concentric circles that close around them or will lie on the edges of the plot. Looking at the graph, we see that $(0, 0)$, $(13, 30)$, $(37, 18)$, and $(32, 34)$ appear to be such points. Since the points near $(0, 0)$ increase in functional value as they close around $(0, 0)$, $f(0, 0)$ will be somewhat more than its nearest contour. So $f(0, 0) \approx 12.5$. Similarly, since the contours near $(0, 0)$ are less in functional value than $f(0, 0)$, $f(0, 0)$ is a local maximum. Applying analogous arguments to the other extreme points, we see that $f(13, 30) \approx 4.5$ and is a local minimum; $f(37, 18) \approx 2.5$ and is a local minimum; and $f(32, 34) \approx 10.5$ and is a local maximum. Since none of the local maxima are greater in value than $f(0, 0) \approx 12.5$, $f(0, 0)$ is a global maximum. Since none of the local minima are lower in value than $f(37, 18) \approx 2.5$, $f(37, 18)$ is a global minimum.

17. At a local maximum value of f,

$$\frac{\partial f}{\partial x} = -2x - B = 0.$$

We are told that this is satisfied by $x = -2$. So $-2(-2) - B = 0$ and $B = 4$. In addition,

$$\frac{\partial f}{\partial y} = -2y - C = 0$$

and we know this holds for $y = 1$, so $-2(1) - C = 0$, giving $C = -2$. We are also told that the value of f is 15 at the point $(-2, 1)$, so

$$15 = f(-2, 1) = A - ((-2)^2 + 4(-2) + 1^2 - 2(1)) = A - (-5), \text{ so } A = 10.$$

Now we check that these values of A, B, and C give $f(x, y)$ a local maximum at the point $(-2, 1)$. Since

$$f_{xx}(-2, 1) = -2,$$
$$f_{yy}(-2, 1) = -2$$

and

$$f_{xy}(-2, 1) = 0,$$

we have that $f_{xx}(-2, 1) f_{yy}(-2, 1) - f_{xy}^2(-2, 1) = (-2)(-2) - 0 > 0$ and $f_{xx}(-2, 1) < 0$. Thus, f has a local maximum value 15 at $(-2, 1)$.

21. We calculate the partial derivatives and set them to zero.

$$\frac{\partial \text{ (range)}}{\partial t} = -10t - 6h + 400 = 0$$
$$\frac{\partial \text{ (range)}}{\partial h} = -6t - 6h + 300 = 0.$$

$$10t + 6h = 400$$
$$6t + 6h = 300$$

solving we obtain

$$4t = 100$$

so

$$t = 25$$

Solving for h, we obtain $6h = 150$, yielding $h = 25$. Since the range is quadratic in h and t, the second derivative test tells us this is a local and global maximum. So the optimal conditions are $h = 25\%$ humidity and $t = 25°$C.

Solutions for Section 9.6

1. We wish to optimize $f(x,y) = xy$ subject to the constraint $g(x,y) = 5x + 2y = 100$. To do this we must solve the following system of equations:

$$\begin{aligned} f_x(x,y) &= \lambda g_x(x,y), \qquad \text{so } y = 5\lambda \\ f_y(x,y) &= \lambda g_y(x,y), \qquad \text{so } x = 2\lambda \\ g(x,y) &= 100, \qquad \text{so } 5x + 2y = 100 \end{aligned}$$

We substitute in the third equation to obtain $5(2\lambda) + 2(5\lambda) = 100$, so $\lambda = 5$. Thus,

$$x = 10 \quad y = 25 \quad \lambda = 5$$

corresponding to optimal $f(x,y) = (10)(25) = 250$.

5. Our objective function is $f(x,y) = x + y$ and our equation of constraint is $g(x,y) = x^2 + y^2 = 1$. To optimize $f(x,y)$ with Lagrange multipliers, we solve the following system of equations

$$\begin{aligned} f_x(x,y) &= \lambda g_x(x,y), \qquad \text{so } 1 = 2\lambda x \\ f_y(x,y) &= \lambda g_y(x,y), \qquad \text{so } 1 = 2\lambda y \\ g(x,y) &= 1, \qquad \text{so } x^2 + y^2 = 1 \end{aligned}$$

Solving for λ gives

$$\lambda = \frac{1}{2x} = \frac{1}{2y},$$

which tells us that $x = y$. Going back to our equation of constraint, we use the substitution $x = y$ to solve for y:

$$\begin{aligned} g(y,y) = y^2 + y^2 &= 1 \\ 2y^2 &= 1 \\ y^2 &= \frac{1}{2} \\ y &= \pm\sqrt{\frac{1}{2}} = \pm\frac{\sqrt{2}}{2}. \end{aligned}$$

Since $x = y$, our critical points are $(\frac{\sqrt{2}}{2}, \frac{\sqrt{2}}{2})$ and $(-\frac{\sqrt{2}}{2}, -\frac{\sqrt{2}}{2})$. Since the constraint is closed and bounded, maximum and minimum values of f subject to the constraint exist. Evaluating f at the critical points we find that the maximum value is $f(\frac{\sqrt{2}}{2}, \frac{\sqrt{2}}{2}) = \sqrt{2}$ and the minimum value is $f(-\frac{\sqrt{2}}{2}, -\frac{\sqrt{2}}{2}) = -\sqrt{2}$.

9. Our objective function is $f(x,y) = xy$ and our equation of constraint is $g(x,y) = 4x^2 + y^2 = 8$. Their partial derivatives are

$$\begin{aligned} f_x &= y, \qquad f_y = x \\ g_x &= 8x, \qquad g_y = 2y. \end{aligned}$$

This gives

$$8x\lambda = y \quad \text{and} \quad 2y\lambda = x.$$

Multiplying, we get

$$8x^2\lambda = 2y^2\lambda.$$

If $\lambda = 0$, then $x = y = 0$, which doesn't satisfy the constraint equation. So $\lambda \neq 0$ and we get

$$\begin{aligned} 2y^2 &= 8x^2 \\ y^2 &= 4x^2 \\ y &= \pm 2x. \end{aligned}$$

To find x, we substitute for y in our equation of constraint.

$$\begin{aligned} 4x^2 + y^2 &= 8 \\ 4x^2 + 4x^2 &= 8 \\ x^2 &= 1 \\ x &= \pm 1 \end{aligned}$$

So our critical points are $(1,2)$, $(1,-2)$, $(-1,2)$ and $(-1,-2)$. Evaluating $f(x,y)$ at the critical points, we have

$$f(1,2) = f(-1,-2) = 2$$
$$f(1,-2) = f(1,-2) = -2.$$

Thus, the maximum value of f on $g(x,y) = 8$ is 2, and the minimum value is -2.

13. **(a)** To be producing the maximum quantity Q under the cost constraint given, the firm should be using K and L values given by

$$\frac{\partial Q}{\partial K} = 0.6aK^{-0.4}L^{0.4} = 20\lambda$$
$$\frac{\partial Q}{\partial L} = 0.4aK^{0.6}L^{-0.6} = 10\lambda$$
$$20K + 10L = 150.$$

Hence $\dfrac{0.6aK^{-0.4}L^{0.4}}{0.4aK^{0.6}L^{-0.6}} = 1.5\dfrac{L}{K} = \dfrac{20\lambda}{10\lambda} = 2$, so $L = \dfrac{4}{3}K$. Substituting in $20K + 10L = 150$, we obtain $20K + 10\left(\dfrac{4}{3}\right)K = 150$. Then $K = \dfrac{9}{2}$ and $L = 6$, so capital should be reduced by $\dfrac{1}{2}$ unit, and labor should be increased by 1 unit.

(b) $\dfrac{\text{New production}}{\text{Old production}} = \dfrac{a4.5^{0.6}6^{0.4}}{a5^{0.6}5^{0.4}} \approx 1.01$, so tell the board of directors, "Reducing the quantity of capital by $1/2$ unit and increasing the quantity of labor by 1 unit will increase production by 1% while holding costs to \$150."

17. The value of λ tells us how the optimum value of f changes when we change the constraint. In particular, if the constraint increases by 1, the optimum value of f increases by λ.

(a) So, if we raise the quota by 1 product, cost rises by $\lambda = 15$. So $C = 1200 + 15 = 1215$.
(b) Similarly, if we lower the quota by 1, cost falls by $\lambda = 15$. So $C = 1200 - 15 = 1185$.

21. **(a)** The curves are shown in Figure 9.4.

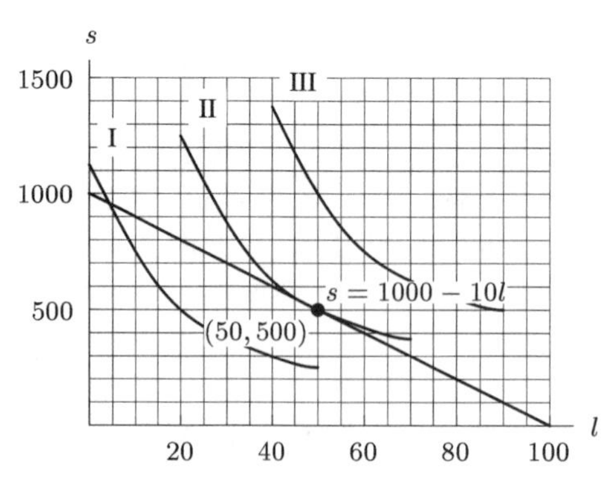

Figure 9.4

(b) The income equals \$10/hour times the number of hours of work:

$$s = 10(100 - l) = 1000 - 10l.$$

(c) The graph of this constraint is the straight line in Figure 9.4.
(d) For any given salary, curve III allows for the most leisure time, curve I the least. Similarly, for any amount of leisure time, curve III also has the greatest salary, and curve I the least. Thus, any point on curve III is preferable to any point on curve II, which is preferable to any point on curve I. We prefer to be on the outermost curve that our constraint allows. We want to choose the point on $s = 1000 - 10l$ which is on the most preferable curve. Since all the curves are concave up, this occurs at the point where $s = 1000 - 10l$ is *tangent* to curve II. So we choose $l = 50$, $s = 500$, and work 50 hours a week.

Solutions for Chapter 9 Review

1. **(a)** The profit is given by the following:

$$\pi = (\text{Revenue from } q_1) + (\text{ Revenue from } q_2) - \text{Cost}.$$

Measuring π in thousands, we obtain:

$$\pi = 3q_1 + 12q_2 - 4.$$

(b) A contour diagram of π is in Figure 9.5. Note that the units of π are in thousands.

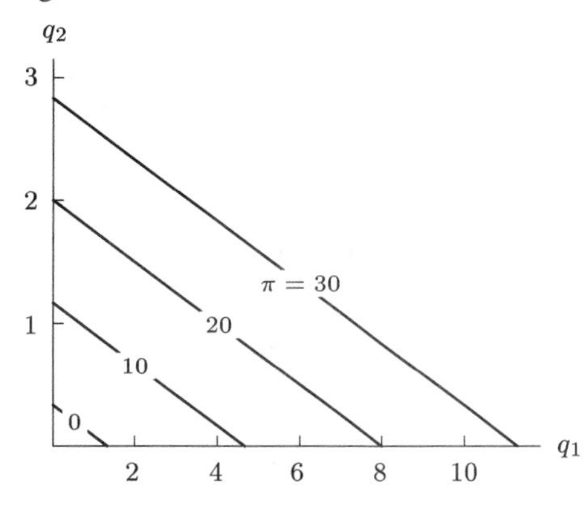

Figure 9.5

5. The contours of $z = y - \sin x$ are of the form $y = \sin x + c$ for a constant c. They are sinusoidal graphs shifted vertically by the value of z on the contour line. The contours are equally spaced vertically for equally spaced z values. See Figure 9.6.

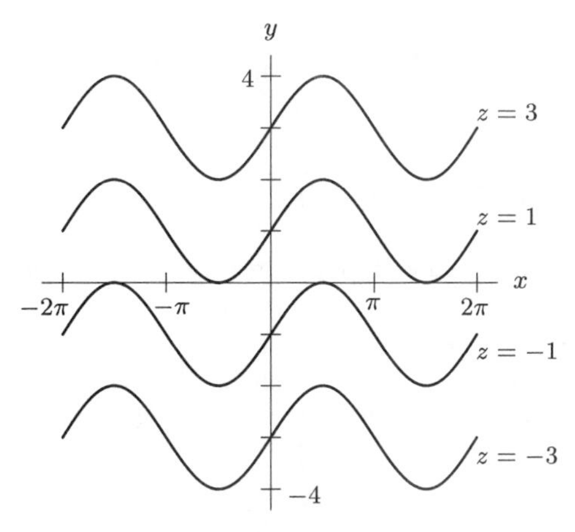

Figure 9.6

9. **(a)** The TMS map of an eye of constant curvature will have only one color, with no contour lines dividing the map.

(b) The contour lines are circles, because the cross-section is the same in every direction. The largest curvature is in the center. See picture below.

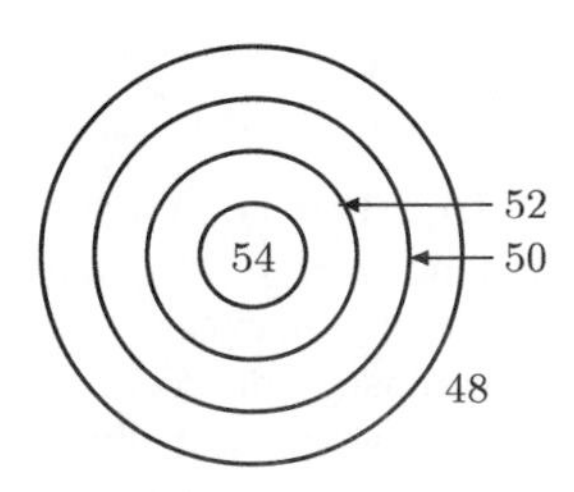

13. $\dfrac{\partial f}{\partial x} = 5e^{-2t}$, $\dfrac{\partial f}{\partial t} = -10xe^{-2t}$.

17. The function $y = f(x, 0) = \cos 0 \sin x = \sin x$ gives the displacement of each point of the string when time is held fixed at $t = 0$. The function $f(x, 1) = \cos 1 \sin x = 0.54 \sin x$ gives the displacement of each point of the string at time $t = 1$. Graphing $f(x, 0)$ and $f(x, 1)$ gives in each case an arch of the sine curve, the first with amplitude 1 and the second with amplitude 0.54. For each different fixed value of t, we get a different snapshot of the string, each one a sine curve with amplitude given by the value of $\cos t$. The result looks like the sequence of snapshots shown in Figure 9.7.

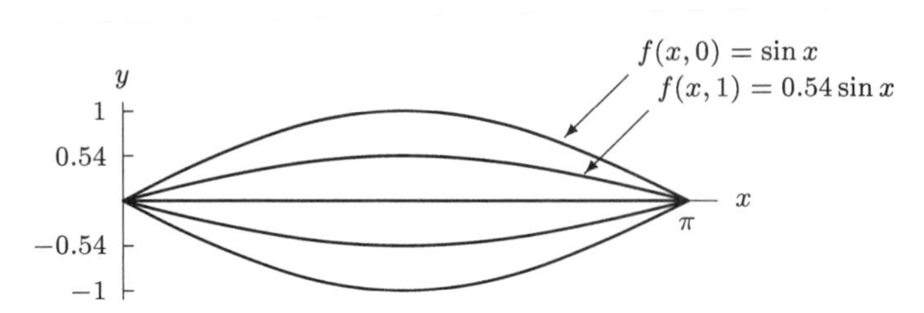

Figure 9.7

21. (a) Estimating $T(x, t)$ from the figure in the text at $x = 15$, $t = 20$ gives

$$\left.\frac{\partial T}{\partial x}\right|_{(15,20)} \approx \frac{T(23,20) - T(15,20)}{23 - 15} = \frac{20 - 23}{8} = -\frac{3}{8}\ ^\circ\text{C per m},$$

$$\left.\frac{\partial T}{\partial t}\right|_{(15,20)} \approx \frac{T(15,25) - T(15,20)}{25 - 20} = \frac{25 - 23}{5} = \frac{2}{5}\ ^\circ\text{C per min}.$$

At 15 m from heater at time $t = 20$ min, the room temperature decreases by approximately $3/8^\circ$C per meter and increases by approximately $2/5^\circ$C per minute.

(b) We have the estimates,

$$\left.\frac{\partial T}{\partial x}\right|_{(5,12)} \approx \frac{T(7,12) - T(5,12)}{7 - 5} = \frac{25 - 27}{2} = -1\ ^\circ\text{C per m},$$

$$\left.\frac{\partial T}{\partial t}\right|_{(5,12)} \approx \frac{T(5,40) - T(5,12)}{40 - 12} = \frac{30 - 27}{28} = \frac{3}{28}\ ^\circ\text{C per min}.$$

At $x = 5$, $t = 12$ the temperature decreases by approximately 1°C per meter and increases by approximately $3/28^\circ$C per minute.

25. **(a)** C represents the quantity of its drinks demanded if $p_1 = p_2 = p_3 = 0$. This represents the quantity of its drinks which would be given away for free when the competing drinks are also free and no money is spent on advertising.

(b) Marginal demand with respect to changes in the company's soft-drink prices is given by

$$\frac{\partial q}{\partial p_1} = -8 \cdot 10^6.$$

Marginal demand with respect to changes in the soft drink prices of other companies is given by

$$\frac{\partial q}{\partial p_2} = 4 \cdot 10^6.$$

Marginal demand with respect to changes in the money spent on advertising is given by

$$\frac{\partial q}{\partial p_3} = 2.$$

The answers are reasonable because the largest effect on quantity sold of the company's soft drinks is caused by changes in their prices and as their prices increase, quantity falls (hence a negative marginal demand). The second largest effect on the quantity sold is caused by the changes in the prices of competing soft drinks. As other companies raise their prices, quantity sold of the first company's drinks increases (hence a positive marginal demand). The effect of changes in money spent on advertising is the smallest and the more money the company spends advertising its drinks, the more drinks it will sell, so there is a positive marginal demand.

29. Since $\frac{\partial f}{\partial x}$ and $\frac{\partial f}{\partial y}$ are defined everywhere, a critical point will occur where $\frac{\partial f}{\partial x} = 0$ and $\frac{\partial f}{\partial y} = 0$. So:

$$\frac{\partial f}{\partial x} = 3x^2 - 3 = 0, \qquad \text{so } x^2 = 1 \text{ and } x = \pm 1$$
$$\frac{\partial f}{\partial y} = 2y = 0, \qquad \text{so } y = 0$$

So $(1, 0)$ and $(-1, 0)$ are the critical points. To determine whether these are local extrema, we can examine values of f for (x, y) near the critical points. Functional values for points near $(1, 0)$ are shown in Table 9.6:

Table 9.6

y \ x	0.99	1.00	1.01
−0.01	−1.9996	−1.9999	−1.9996
0.00	−1.9997	−2.0000	−1.9997
0.01	−1.9996	−1.9999	−1.9996

As we can see from the table, all the points close to $(1, 0)$ have greater functional value, so $f(1, 0)$ is a local minimum. A similar display of points near $(-1, 0)$ is shown in Table 9.7:

Table 9.7

y \ x	−1.01	−1.00	−0.99
−0.01	1.9998	2.0001	1.9998
0.00	1.9997	2.0000	1.9997
0.01	1.9998	2.0001	1.9998

As we can see, some points near $(-1, 0)$ have greater functional value than $f(-1, 0)$ and others have less. So $f(-1, 0)$ is neither a local maximum nor a local minimum.

33. **(a)** Because of budget constraints, we are limited in the size of the labor force and the amount of total equipment. This constraint is described in the formula

$$300L + 100K = 15{,}000$$

We can let L range from 0 to about 50. Each choice of L determines a choice of K. We have, as a result, the Table 9.8.

Table 9.8

L	5	10	15	20	25	30	35	40	45	50
K	135	120	105	90	75	60	45	30	15	0
P	48	82	111	135	156	172	184	189	181	0

(b) We wish to maximize P subject to that cost C satisfies $C = 300L + 100K = \$15{,}000$. This is accomplished by solving the following system of equations:

$$P_L = \lambda C_L, \qquad \text{so } 4L^{-0.2}K^{0.2} = 300\lambda$$
$$P_K = \lambda C_K, \qquad \text{so } L^{0.8}K^{-0.8} = 100\lambda$$
$$C = 15{,}000, \qquad \text{so } 300L + 100K = 15{,}000$$

We can solve these equations by dividing the first by the second and then substituting into the budget constraint. Doing so produces the solution $K = 30$, $L = 40$. So the optimal choices of labor and capital are $L = 40$, $K = 30$.

Solutions to Problems on Deriving the Formula for Regression Lines

1. Let the line be in the form $y = b + mx$. When x equals -1, 0 and 1, then y equals $b - m$, b, and $b + m$, respectively. The sum of the squares of the vertical distances, which is what we want to minimize, is

$$f(m, b) = (2 - (b - m))^2 + (-1 - b)^2 + (1 - (b + m))^2.$$

To find the critical points, we compute the partial derivatives with respect to m and b,

$$\begin{aligned} f_m &= 2(2 - b + m) + 0 + 2(1 - b - m)(-1) \\ &= 4 - 2b + 2m - 2 + 2b + 2m \\ &= 2 + 4m, \\ f_b &= 2(2 - b + m)(-1) + 2(-1 - b)(-1) + 2(1 - b - m)(-1) \\ &= -4 + 2b - 2m + 2 + 2b - 2 + 2b + 2m \\ &= -4 + 6b. \end{aligned}$$

Setting both partial derivatives equal to zero, we get a system of equations:

$$\begin{aligned} 2 + 4m &= 0, \\ -4 + 6b &= 0. \end{aligned}$$

The solution is $m = -1/2$ and $b = 2/3$. You can check that it is a minimum. Hence, the regression line is $y = \frac{2}{3} - \frac{1}{2}x$.

5. We have $\sum x_i = 6$, $\sum y_i = 5$, $\sum x_i^2 = 14$, and $\sum y_i x_i = 12$. Thus

$$\begin{aligned} b &= (14 \cdot 5 - 6 \cdot 12) \,/\, \left(3 \cdot 14 - 6^2\right) = -1/3. \\ m &= (3 \cdot 12 - 6 \cdot 5) \,/\, \left(3 \cdot 14 - 6^2\right) = 1. \end{aligned}$$

The line is $y = x - \frac{1}{3}$, which agrees with the answer to Example 1.

CHAPTER TEN

Solutions for Section 10.1

1. **(a)** (III) An island can only sustain the population up to a certain size. The population will grow until it reaches this limiting value.
   (b) (V) The ingot will get hot and then cool off, so the temperature will increase and then decrease.
   (c) (I) The speed of the car is constant, and then decreases linearly when the breaks are applied uniformly.
   (d) (II) Carbon-14 decays exponentially.
   (e) (IV) Tree pollen is seasonal, and therefore cyclical.

5. **(a)** The amount of caffeine, A, is decreasing at a rate of 17% times A, so we have

$$\frac{dA}{dt} = -0.17A.$$

   The negative sign indicates that the amount of caffeine is decreasing at a rate of 17% times A. Notice that the initial amount of caffeine, 100 mg, is not used in the differential equation. The differential equation tells us only how things are changing.
   (b) At the start of the first hour, we have $A = 100$. Substituting this into the differential equation, we have

$$\frac{dA}{dt} = -0.17A = -0.17(100) = -17 \text{ mg/hour}.$$

   We estimate that the amount of caffeine decreases by about $(17 \text{ mg/hr}) \cdot (1 \text{ hr}) = 17$ mg during the first hour. This is only an estimate, however, since the derivative dA/dt will not stay constant at -17 throughout the entire first hour.

9. The amount of toxin, A, is increasing at a rate of 10 micrograms per day and is decreasing at a rate of 0.03 times A. We have

$$\begin{aligned} \text{Rate of change of } A &= \text{Rate in} - \text{Rate out.} \\ \frac{dA}{dt} &= 10 - 0.03A. \end{aligned}$$

13. The quantity y is increasing when dy/dt is positive. Using the differential equation, we see that dy/dt is positive when $-0.5y$ is positive, which means y is negative. The quantity y is increasing when y is negative. Similarly, the quantity y is decreasing when y is positive.

17. Since $dy/dx = 2$, the slope of the solution curve will be 2 at all points. Any possible solution curve for this differential equation will be a line with slope 2. A possible solution curve for this differential equation is Graph F.

Solutions for Section 10.2

1. Since $y = t^4$, the derivative is $dy/dt = 4t^3$. We have

$$\begin{aligned} \text{Left-side} &= t\frac{dy}{dt} = t(4t^3) = 4t^4. \\ \text{Right-side} &= 4y = 4t^4. \end{aligned}$$

   Since the substitution $y = t^4$ makes the differential equation true, $y = t^4$ is in fact a solution.

5. We know that at time $t = 0$, the value of y is 8. Since we are told that $dy/dt = 0.5y$, we know that at time $t = 0$

$$\frac{dy}{dt} = 0.5(8) = 4.$$

As t goes from 0 to 1, y will increase by 4, so at $t = 1$,

$$y = 8 + 4 = 12.$$

Likewise, we get that at $t = 1$,

$$\frac{dy}{dt} = .5(12) = 6$$

so that at $t = 2$,

$$y = 12 + 6 = 18.$$

At $t = 2$, $\frac{dy}{dt} = .5(18) = 9$ so that at $t = 3$, $y = 18 + 9 = 27$.
At $t = 3$, $\frac{dy}{dt} = .5(27) = 13.5$ so that at $t = 4$, $y = 27 + 13.5 = 40.5$.
Thus we get the following table

t	0	1	2	3	4
y	8	12	18	27	40.5

9. (a) = (III), (b) = (IV), (c) = (I), (d) = (II).

13. Since $y = x^2 + k$ we know that

$$y' = 2x.$$

Substituting $y = x^2 + k$ and $y' = 2x$ into the differential equation we get

$$\begin{aligned} 10 &= 2y - xy' \\ &= 2(x^2 + k) - x(2x) \\ &= 2x^2 + 2k - 2x^2 \\ &= 2k \end{aligned}$$

Thus, $k = 5$ is the only solution.

17. We first compute dy/dx for each of the functions on the right.

If $y = x^3$ then

$$\begin{aligned} \frac{dy}{dx} &= 3x^2 \\ &= 3\frac{y}{x}. \end{aligned}$$

If $y = 3x$ then

$$\begin{aligned} \frac{dy}{dx} &= 3 \\ &= \frac{y}{x}. \end{aligned}$$

If $y = e^{3x}$ then

$$\begin{aligned} \frac{dy}{dx} &= 3e^{3x} \\ &= 3y. \end{aligned}$$

If $y = 3e^x$ then

$$\begin{aligned} \frac{dy}{dx} &= 3e^x \\ &= y. \end{aligned}$$

Finally, if $y = x$ then

$$\frac{dy}{dx} = 1$$
$$= \frac{y}{x}.$$

Comparing our calculated derivatives with the right-hand sides of the differential equations we see that (a) is solved by (II) and (V), (b) is solved by (I), (c) is not solved by any of our functions, (d) is solved by (IV) and (e) is solved by (III).

Solutions for Section 10.3

1.

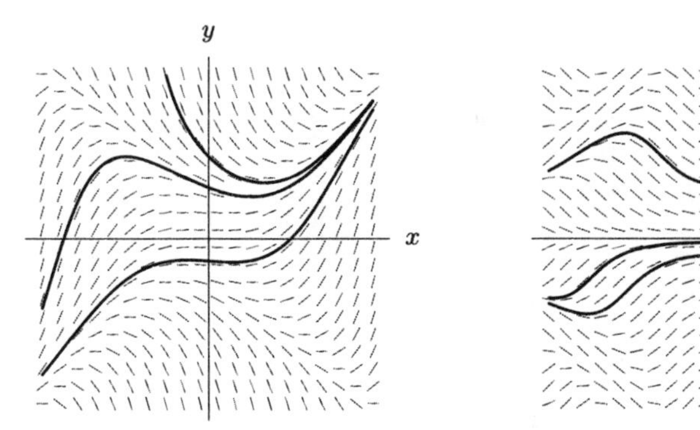

Figure 10.1

Other choices of solution curves are, of course, possible.

5. **(a)** The slope at any point (x, y) is equal to dy/dx, which is xy.Then:

$$\begin{aligned} &\text{At point } (2,1), \text{ slope } = 2 \cdot 1 = 2, \\ &\text{At point } (0,2), \text{ slope } = 0 \cdot 2 = 0, \\ &\text{At point } (-1,1), \text{ slope } = -1 \cdot 1 = -1, \\ &\text{At point } (2,-2), \text{ slope } = 2(-2) = -4. \end{aligned}$$

(b)

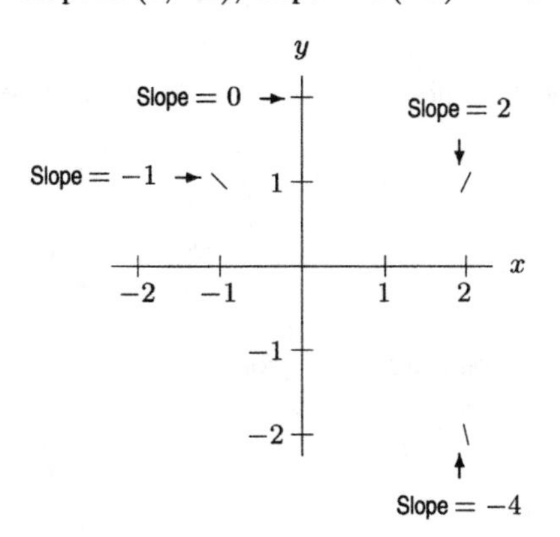

Figure 10.2

9. As x increases, $y \to \infty$.

13. From the slope field, the function looks like a parabola of the form $y = x^2 + C$, where C depends on the starting point. In any case, $y \to \infty$ as $x \to \infty$.

Solutions for Section 10.4

1. The equation given is in the form

$$\frac{dP}{dt} = kP.$$

Thus we know that the general solution to this equation will be

$$P = Ce^{kt}.$$

And in our case, with $k = 0.02$ and $C = 20$ we get

$$P = 20e^{0.02t}.$$

5. The equation given is in the form

$$\frac{dQ}{dt} = kQ.$$

Thus we know that the general solution to this equation will be

$$Q = Ce^{kt}.$$

And in our case, with $k = \frac{1}{5}$ we get

$$Q = Ce^{\frac{1}{5}t}.$$

We know that $Q = 50$ when $t = 0$. Thus we get

$$\begin{aligned} Q(t) &= Ce^{\frac{1}{5}t} \\ Q(0) &= 50 = Ce^{0} \\ 50 &= C \end{aligned}$$

Thus we get

$$Q = 50e^{\frac{1}{5}t}.$$

9. **(a)** Since the growth rate of the tumor is proportional to its size, we should have

$$\frac{dS}{dt} = kS.$$

(b) We can solve this differential equation by separating variables and then integrating:

$$\begin{aligned} \int \frac{dS}{S} &= \int k\,dt \\ \ln|S| &= kt + B \\ S &= Ce^{kt}. \end{aligned}$$

(c) This information is enough to allow us to solve for C:

$$\begin{aligned} 5 &= Ce^{0t} \\ C &= 5. \end{aligned}$$

(d) Knowing that $C = 5$, this second piece of information allows us to solve for k:

$$\begin{aligned} 8 &= 5e^{3k} \\ k &= \frac{1}{3}\ln\left(\frac{8}{5}\right) \approx 0.1567. \end{aligned}$$

So the tumor's size is given by

$$S = 5e^{0.1567t}.$$

13. Lake Superior will take the longest, because the lake is largest (V is largest) and water is moving through it most slowly (r is smallest). Lake Erie looks as though it will take the least time because V is smallest and r is close to the largest. For Erie, $k = r/V = 175/460 \approx 0.38$. The lake with the largest value of r is Ontario, where $k = r/V = 209/1600 \approx 0.13$. Since e^{-kt} decreases faster for larger k, Lake Erie will take the shortest time for any fixed fraction of the pollution to be removed.

For Lake Superior,

$$\frac{dQ}{dt} = -\frac{r}{V}Q = -\frac{65.2}{12{,}200}Q \approx -0.0053Q$$

so

$$Q = Q_0 e^{-0.0053t}.$$

When 80% of the pollution has been removed, 20% remains so $Q = 0.2Q_0$. Substituting gives us

$$0.2Q_0 = Q_0 e^{-0.0053t}$$

so

$$t = -\frac{\ln(0.2)}{0.0053} \approx 301 \text{ years}.$$

(Note: The 301 is obtained by using the exact value of $\frac{r}{V} = \frac{65.2}{12{,}200}$, rather than 0.0053. Using 0.0053 gives 304 years.)
For Lake Erie, as in the text

$$\frac{dQ}{dt} = -\frac{r}{V}Q = -\frac{175}{460}Q \approx -0.38Q$$

so

$$Q = Q_0 e^{-0.38t}.$$

When 80% of the pollution has been removed

$$0.2Q_0 = Q_0 e^{-0.38t}$$
$$t = -\frac{\ln(0.2)}{0.38} \approx 4 \text{ years}.$$

So the ratio is

$$\frac{\text{Time for Lake Superior}}{\text{Time for Lake Erie}} \approx \frac{301}{4} \approx 75.$$

In other words it will take about 75 times as long to clean Lake Superior as Lake Erie.

17. **(a)** Suppose $Y(t)$ is the quantity of oil in the well at time t. We know that the oil in the well decreases at a rate proportional to $Y(t)$, so

$$\frac{dY}{dt} = -kY.$$

Integrating, and using the fact that initially $Y = Y_0 = 10^6$, we have

$$Y = Y_0 e^{-kt} = 10^6 e^{-kt}.$$

In six years, $Y = 500,000 = 5 \cdot 10^5$, so

$$5 \cdot 10^5 = 10^6 e^{-k \cdot 6}$$

so

$$0.5 = e^{-6k}$$
$$k = -\frac{\ln 0.5}{6} = 0.1155.$$

When $Y = 600,000 = 6 \cdot 10^5$,

$$\text{Rate at which oil decreasing} = \left|\frac{dY}{dt}\right| = kY = 0.1155(6 \cdot 10^5) = 69{,}300 \text{ barrels/year}.$$

(b) We solve the equation

$$5 \cdot 10^4 = 10^6 e^{-0.1155t}$$
$$0.05 = e^{-0.1155t}$$
$$t = \frac{\ln 0.05}{-0.1155} = 25.9 \text{ years}.$$

Solutions for Section 10.5

1. We know that the general solution to a differential equation of the form

$$\frac{dH}{dt} = k(H - A)$$

is

$$H = A + Ce^{kt}.$$

Thus in our case we get

$$H = 75 + Ce^{3t}.$$

We know that at $t = 0$ we have $H = 0$, so solving for C we get

$$\begin{aligned} H &= 75 + Ce^{3t} \\ 0 &= 75 + Ce^{3(0)} \\ -75 &= Ce^{0} \\ C &= -75. \end{aligned}$$

Thus we get

$$H = 75 - 75e^{3t}.$$

5. We know that the general solution to a differential equation of the form

$$\frac{dQ}{dt} = k(Q - A)$$

is

$$H = A + Ce^{kt}.$$

To get our equation in this form we factor out a 0.3 to get

$$\frac{dQ}{dt} = 0.3\left(Q - \frac{120}{0.3}\right) = 0.3(Q - 400).$$

Thus in our case we get

$$Q = 400 + Ce^{0.3t}.$$

We know that at $t = 0$ we have $Q = 50$, so solving for C we get

$$\begin{aligned} Q &= 400 + Ce^{0.3t} \\ 50 &= 400 + Ce^{0.3(0)} \\ -350 &= Ce^{0} \\ C &= -350. \end{aligned}$$

Thus we get

$$Q = 400 - 350e^{0.3t}.$$

9. In order to check that $y = A + Ce^{kt}$ is a solution to the differential equation

$$\frac{dy}{dt} = k(y - A),$$

we must show that the derivative of y with respect to t is equal to $k(y - A)$:

$$\begin{aligned} y &= A + Ce^{kt} \\ \frac{dy}{dt} &= 0 + (Ce^{kt})(k) \\ &= kCe^{kt} \\ &= k(Ce^{kt} + A - A) \\ &= k\left((Ce^{kt} + A) - A\right) \\ &= k(y - A) \end{aligned}$$

13. Let $D(t)$ be the quantity of dead leaves, in grams per square centimeter. Then $\frac{dD}{dt} = 3 - 0.75D$, where t is in years. We know that the general solution to a differential equation of the form

$$\frac{dD}{dt} = k(D - B)$$

is

$$D = Ae^{kt} + B.$$

Factoring out a -0.75 on the left side we get

$$\frac{dD}{dt} = -0.75\left(D - \frac{-3}{-0.75}\right) = -0.75(D - 4).$$

Thus in our case we get

$$D = Ae^{-0.75t} + 4.$$

If initially the ground is clear, the solution looks like the following graph:

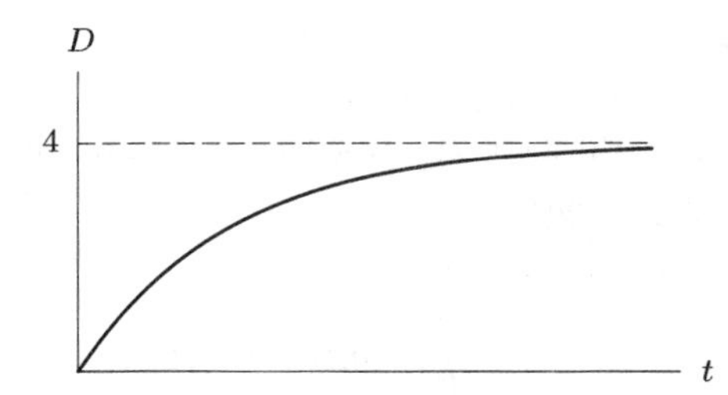

The equilibrium level is 4 grams per square centimeter, regardless of the initial condition.

17. **(a)** We know that the equilibrium solution is the solution satisfying the differential equation whose derivative is everywhere 0. Thus we must solve

$$\frac{dy}{dt} = 0.$$

Solving this gives

$$\begin{aligned} \frac{dy}{dt} &= 0 \\ 0.5y - 250 &= 0 \\ y &= 500 \end{aligned}$$

(b) We know that the general solution to a differential equation of the form

$$\frac{dy}{dt} = k(y - A)$$

is

$$y = A + Ce^{kt}.$$

To get our equation in this form we factor out a 0.5 to get

$$\frac{dy}{dt} = 0.5\left(y - \frac{250}{0.5}\right) = 0.5(y - 500).$$

Thus in our case we get

$$y = 500 + Ce^{0.5t}.$$

(c) The graphs of several solutions is shown in Figure 10.3.

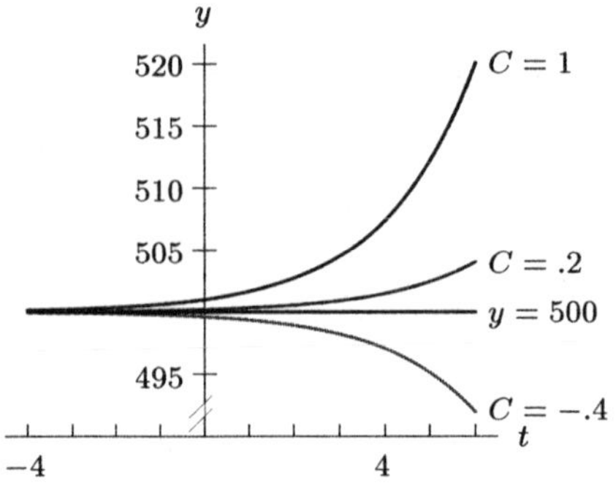

Figure 10.3

(d) Looking at Figure 10.3 we see that as $t \to \infty$, the value of y gets further and further away from the line $y = 500$. The equilibrium solution $y = 500$ is unstable.

21. **(a)** If $B = f(t)$ (where t is in years)

$$\begin{aligned}\frac{dB}{dt} &= \text{Rate at which interest is earned} + \text{Rate at which money is deposited}\\ &= 0.10B + 1000.\end{aligned}$$

(b)

$$\frac{dB}{dt} = 0.1(B + 10{,}000)$$

We know that a differential equation of the form

$$\frac{dB}{dt} = k(B - A)$$

has general solution:

$$B = Ce^{kt} + A.$$

Thus, in our case

$$B = Ce^{0.1t} - 10{,}000.$$

For $t = 0$, $B = 0$, hence $C = 10{,}000$. Therefore, $B = 10{,}000e^{0.1t} - 10{,}000$.

25. **(a)** $\frac{dy}{dt} = -k(y - a)$, where $k > 0$ and a are constants.

(b) We know that the general solution to a differential equation of the form

$$\frac{dy}{dt} = -k(y - a)$$

is

$$y = Ce^{-kt} + a.$$

We can assume that right after the course is over (at $t = 0$) 100% of the material is remembered. Thus we get

$$\begin{aligned}y &= Ce^{-kt} + a\\ 1 &= Ce^{0} + a\\ C &= 1 - a.\end{aligned}$$

Thus

$$y = (1 - a)e^{-kt} + a.$$

(c) As $t \to \infty$, $e^{-kt} \to 0$, so $y \to a$.
Thus, a represents the fraction of material which is remembered in the long run. The constant k tells us about the rate at which material is forgotten.

Solutions for Section 10.6

1. Here x and y both increase at about the same rate.

5. This is an example of a predator-prey relationship. Normally, we would expect the worm population, in the absence of predators, to increase without bound. As the number of worms w increases, so would the rate of increase dw/dt; in other words, the relation $dw/dt = w$ might be a reasonable model for the worm population in the absence of predators.

   However, since there are predators (robins), dw/dt won't be that big. We must lessen dw/dt. It makes sense that the more interaction there is between robins and worms, the more slowly the worms are able to increase their numbers. Hence we lessen dw/dt by the amount wr to get $dw/dt = w - wr$. The term $-wr$ reflects the fact that more interactions between the species means slower reproduction for the worms.

   Similarly, we would expect the robin population to decrease in the absence of worms. We'd expect the population decrease at a rate related to the current population, making $dr/dt = -r$ a reasonable model for the robin population in absence of worms. The negative term reflects the fact that the greater the population of robins, the more quickly they are dying off. The wr term in $dr/dt = -r + wr$ reflects the fact that the more interactions between robins and worms, the greater the tendency for the robins to increase in population.

9. Sketching the trajectory through the point $(2, 2)$ on the slope field given shows that the maximum robin population is about 2500, and the minimum robin population is about 500. When the robin population is at its maximum, the worm population is about 1,000,000.

13. **(a)** Substituting $w = 2.2$ and $r = 1$ into the differential equations gives

$$\frac{dw}{dt} = 2.2 - (2.2)(1) = 0$$
$$\frac{dr}{dt} = -1 + 1(2.2) = 1.2.$$

(b) Since the rate of change of w with time is 0,

$$\text{At } t = 0.1, \quad \text{we estimate } w = 2.2$$

Since the rate of change of r is 1.2 thousand robins per unit time,

$$\text{At } t = 0.1, \quad \text{we estimate } r = 1 + 1.2(0.1) = 1.12 \approx 1.1.$$

(c) We must recompute the derivatives. At $t = 0.1$, we have

$$\frac{dw}{dt} = 2.2 - 2.2(1.12) = -0.264$$
$$\frac{dr}{dt} = -1.12 + 1.12(2.2) = 1.344.$$

Then at $t = 0.2$, we estimate

$$w = 2.2 - 0.264(0.1) = 2.1736 \approx 2.2$$
$$r = 1.12 + 1.344(0.1) = 1.2544 \approx 1.3$$

Recomputing the derivatives at $t = 0.2$ gives

$$\frac{dw}{dt} = 2.1736 - 2.1736(1.2544) = -0.553$$
$$\frac{dr}{dt} = -1.2544 + 1.2544(2.1736) = 1.472$$

Then at $t = 0.3$, we estimate

$$w = 2.1736 - 0.553(0.1) = 2.1183 \approx 2.1$$
$$r = 1.2544 + 1.472(0.1) = 1.4016 \approx 1.4.$$

17. $\dfrac{dx}{dt} = -x + xy, \quad \dfrac{dy}{dt} = y$

21. **(a)** The x population is unaffected by the y population—it grows exponentially no matter what the y population is, even if $y = 0$. If alone, the y population decreases to zero exponentially, because its equation becomes $dy/dt = -0.1y$.
(b) Here, interaction between the two populations helps the y population but doesn't effect the x population. This is not a predator-prey relationship; instead, this is a one-way relationship, where the y population is helped by the existence of x's. These equations could, for instance, model the interaction of rhinocerouses (x) and dung beetles (y).

Solutions for Section 10.7

1. Susceptible people are infected at a rate proportional to the product of S and I. As susceptible people become infected, S decreases at a rate of aSI and (since these same people are now infected) I increases at the same rate. At the same time, infected people are recovering at a rate proportional to the number infected, so I is decreasing at a rate of bI.

5. **(a)** $I_0 = 1$, $S_0 = 349$
(b) Since $\frac{dI}{dt} = 0.0026SI - 0.5I = 0.0026(349)(1) - 0.5(1) > 0$, so I is increasing. The number of infected people will increase, and the disease will spread. This is an epidemic.

9. The threshold value of S is the value at which I is a maximum. When I is a maximum,

$$\frac{dI}{dt} = 0.04SI - 0.2I = 0,$$

so

$$S = 0.2/0.04 = 5.$$

Solutions for Chapter 10 Review

1. **(a)** (i) If $y = Cx^2$, then $\frac{dy}{dx} = C(2x) = 2Cx$. We have

$$x\frac{dy}{dx} = x(2Cx) = 2Cx^2$$

and

$$3y = 3(Cx^2) = 3Cx^2$$

Since $x\frac{dy}{dx} \neq 3y$, this is not a solution.

(ii) If $y = Cx^3$, then $\frac{dy}{dx} = C(3x^2) = 3Cx^2$. We have

$$x\frac{dy}{dx} = x(3Cx^2) = 3Cx^3,$$

and

$$3y = 3Cx^3.$$

Thus $x\frac{dy}{dx} = 3y$, and $y = Cx^3$ is a solution.

(iii) If $y = x^3 + C$, then $\frac{dy}{dx} = 3x^2$. We have

$$x\frac{dy}{dx} = x(3x^2) = 3x^3$$

and

$$3y = 3(x^3 + C) = 3x^3 + 3C.$$

Since $x\frac{dy}{dx} \neq 3y$, this is not a solution.

(b) The solution is $y = Cx^3$. If $y = 40$ when $x = 2$, we have

$$40 = C(2^3)$$
$$40 = C \cdot 8$$
$$C = 5.$$

5. The general solution is
$$y = Ce^{5t}.$$

9. Multiplying both sides by Q gives
$$\frac{dQ}{dt} = 2Q, \quad \text{where } Q \neq 0.$$
We know that the general solution to an equation of the form
$$\frac{dQ}{dt} = kQ$$
is
$$Q = Ce^{kt}.$$
Thus in our case the solution is
$$Q = Ce^{2t},$$
where C is some constant, $C \neq 0$.

13. We know that the general solution to the differential equation
$$\frac{dH}{dt} = k(H - A)$$
is
$$H = Ce^{kt} + A.$$
Thus in our case we factor out 0.5 to get
$$\frac{dH}{dt} = 0.5\left(H + \frac{10}{0.5}\right) = 0.5(H - (-20)).$$
Thus the general solution to our differential equation is
$$H = Ce^{0.5t} - 20,$$
where C is some constant.

17. We know that the general solution to the differential equation
$$\frac{dH}{dt} = k(H - A)$$
is
$$H = Ce^{kt} + A.$$
Thus in our case we factor out -0.5 to get
$$\frac{dH}{dt} = -0.5\left(H - \frac{100}{-0.5}\right) = -0.5(H - (-200)).$$
Thus the general solution to our differential equation is
$$H = Ce^{-0.5t} - 200.$$
Solving for C with $H(0) = 40$ we get
$$\begin{aligned} H(t) &= Ce^{-0.5t} - 200 \\ 40 &= Ce^{0} - 200 \\ C &= 240. \end{aligned}$$
Thus the solution is
$$H = 240e^{-0.5t} - 200.$$
The graph of this function is shown in Figure 10.4

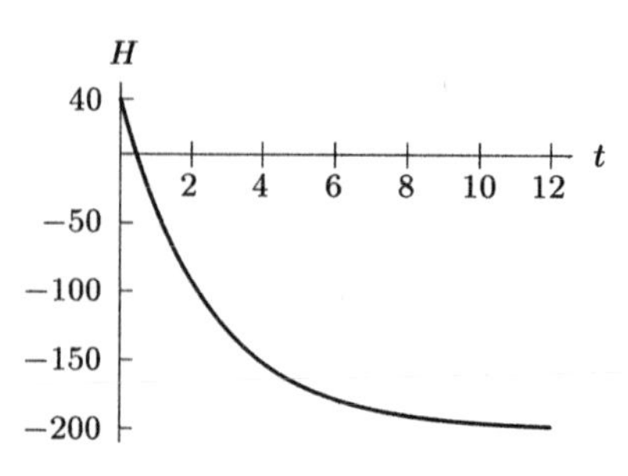

Figure 10.4

21. **(a)** The value of the company satisfies

$$\text{Rate of change of value} = \text{Rate interest earned} - \text{Rate expenses paid}$$

so

$$\frac{dV}{dt} = 0.02V - 80{,}000.$$

(b) We find V when

$$\begin{aligned} \frac{dV}{dt} &= 0 \\ 0.02V - 80{,}000 &= 0 \\ 0.02V &= 80{,}000 \\ V &= 4{,}000{,}000 \end{aligned}$$

There is an equilibrium solution at $V = \$4{,}000{,}000$. If the company has \$4,000,000 in assets, its earnings will exactly equal its expenses.

(c) The general solution is

$$V = 4{,}000{,}000 + Ce^{0.02t}.$$

(d) If $V = 3{,}000{,}000$ when $t = 0$, we have $C = -1{,}000{,}000$. The solution is

$$V = 4{,}000{,}000 - 1{,}000{,}000e^{0.02t}.$$

When $t = 12$, we have

$$\begin{aligned} V &= 4{,}000{,}000 - 1{,}000{,}000e^{0.02(12)} \\ &= 4{,}000{,}000 - 1{,}271{,}249 \\ &= \$2{,}728{,}751. \end{aligned}$$

The company is losing money.

25. Using (Rate balance changes) = (Rate interest is added)– (Rate payments are made), when the interest rate is i, we have

$$\frac{dB}{dt} = iB - 100.$$

We know that the general solution to a differential equation of the form

$$\frac{dB}{dt} = k(B - A)$$

is

$$B = Ce^{kt} + A.$$

Factoring out an i on the left side we get

$$\frac{dB}{dt} = i\left(B - \frac{100}{i}\right).$$

Thus in our case we get

$$B = Ce^{it} + \frac{100}{i}.$$

We know that at $t = 0$ we have $B = 1000$ so solving for C we get

$$1000 = Ce^0 + \frac{100}{i}$$
$$C = 1000 - \frac{100}{i}.$$

Thus $B = \frac{100}{i} + (1000 - \frac{100}{i})e^{it}$.
When $i = 0.05$, $B = 2000 - 1000e^{0.05t}$.
When $i = 0.1$, $B = 1000$.
When $i = 0.15$, $B = 666.67 + 333.33e^{0.15t}$.
We now look at the graph when $i = 0.05$, $i = 0.1$, and $i = 0.15$.

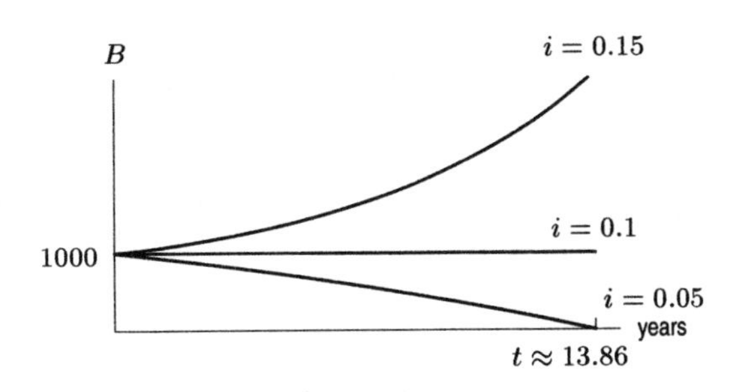

29. **(a)** The species need each other to survive. Both would die out without the other, and they help each other.
(b) If $x = 2$ and $y = 1$,

$$\frac{dx}{dt} = -3x + 2xy = -3(2) + 2(2)(1) < 0,$$

and so population x decreases. If $x = 2$ and $y = 1$,

$$\frac{dy}{dx} = -y + 5xy = -1 + 5(2)(1) > 0,$$

and so population y increases.
(c) $\dfrac{dy}{dx} = \dfrac{-y + 5xy}{-3x + 2xy}$.
(d) See Figure 10.5.
(e) See Figure 10.5.

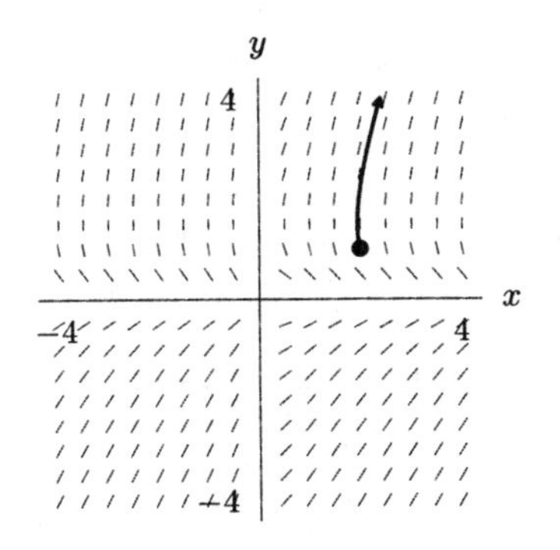

Figure 10.5

CHAPTER ELEVEN

Solutions for Section 11.1

1. Adding the terms, we see that

$$50 + 50(0.9) + 50(0.9)^2 + 50(0.9)^3 = 50 + 45 + 40.5 + 36.45 = 171.95.$$

We can also find the sum using the formula for a finite geometric series with $a = 50$, $r = 0.9$, and $n = 4$:

$$50 + 50(0.9) + 50(0.9)^2 + 50(0.9)^3 = \frac{50(1 - (0.9)^4)}{1 - 0.9} = 171.95.$$

5. This is an infinite geometric series with $a = 30$ and $r = 0.85$. Since $-1 < r < 1$, the series converges and we have

$$\text{Sum} = \frac{30}{1 - 0.85} = 200.$$

9. Each term in this series is half the preceding term, so this is an infinite geometric series with $a = 200$ and $r = 0.5$. Since $-1 < r < 1$, the series converges and we have

$$\text{Sum} = \frac{200}{1 - 0.5} = 400.$$

13. This is a finite geometric series with $a = 3$, $r = 1/2$, and $n - 1 = 10$, so $n = 11$. Thus

$$\text{Sum} = \frac{3(1 - (1/2)^{11})}{1 - 1/2} = 3 \cdot 2\frac{(2^{11} - 1)}{2^{11}} = \frac{3(2^{11} - 1)}{2^{10}}.$$

17. Right after a dose is given, the amount of the drug in the body is the amount from that dose (100 mg) plus the amounts remaining from all previous doses. In the long term

$$\text{Amount of drug in the body after a dose} = 100 + 100(0.82) + 100(0.82)^2 + \cdots.$$

This is an infinite geometric series with $a = 100$ and $r = 0.82$. Since $-1 < r < 1$, the series converges. Using the formula for the sum:

$$\begin{aligned}\text{Long term amount, right after a dose} &= 100 + 100(0.82) + 100(0.82)^2 + \cdots \\ &= \frac{100}{1 - 0.82} = 555.556 \text{ mg.}\end{aligned}$$

Since each dose is 100 mg, the long term amount right before a dose is 100 mg less than the long term amount right after a dose. We have

$$\text{Long term amount, right before a dose} = 555.556 - 100 = 455.556 \text{ mg.}$$

Solutions for Section 11.2

1. The 20th deposit has just been made and has not yet earned any interest, the 19th deposit has earned interest for one year and is worth $1000(1.085)$, the 18th deposit has earned interest for two years and is worth $1000(1.085)^2$, and so on. The first deposit has earned interest for 19 years and is worth $1000(1.085)^{19}$. The total balance in the account right after the 20th deposit is

$$\text{Balance} = 1000 + 1000(1.085) + 1000(1.085)^2 + \cdots + 1000(1.085)^{19}.$$

This is a finite geometric series with $a = 1000$, $r = 1.085$, and $n = 20$. Using the formula for the sum of a finite geometric series, we have

$$\text{Balance after the } 20^{\text{th}} \text{ deposit} = \frac{1000(1 - (1.085)^{20})}{1 - 1.085} = \$48{,}377.01.$$

The twenty annual deposits of \$1000 have contributed a total of \$20,000 to this balance, and the remaining \$28,377.01 in the account comes from the interest earned.

5. **(a)** The present value of the payment made one year from now is $50{,}000(1.072^{-1})$, and the present value of the payment made n years from now is $50{,}000(1.072^{-1})^n$. Since the first payment is made now, the 10th payment is made 9 years from now. Summing, we have

$$\text{Present value } = 50000 + 50000(1.072^{-1}) + 50000(1.072^{-1})^2 + \cdots + 50000(1.072^{-1})^9.$$

This is a finite geometric series with $a = 50000$, $r = 1.072^{-1}$, and $n = 10$. Using the formula for the sum, we have

$$\text{Present value of the annuity } = \frac{50000(1-(1.072^{-1})^{10})}{1-1.072^{-1}} = \$373{,}008.06.$$

(b) If the payments are to continue indefinitely, the present value is the sum:

$$\text{Present value } = 50000 + 50000(1.072^{-1}) + 50000(1.072^{-1})^2 + 50000(1.072^{-1})^3 + \cdots.$$

This is an infinite geometric series with $a = 50000$ and $r = 1.072^{-1} = 0.9328358$. Since $-1 < r < 1$, this series converges and we use the formula for the sum of an infinite geometric series:

$$\text{Present value of the annuity } = \frac{50000}{1-1.072^{-1}} = \$744{,}444.44.$$

9. **(a)** The people who receive the N dollars in tax rebates spend $N \cdot k$ of it. The people who receive this money spend $N \cdot k^2$ of it, and so on. We have

$$\text{Total amount spent } = Nk + Nk^2 + Nk^3 + \cdots.$$

The total amount spent is an infinite geometric series with $a = Nk$ and $r = k$. Since $0 < k < 1$, the series converges. We use the formula for the sum to see

$$\text{Total amount spent } = \frac{Nk}{1-k} = N\left(\frac{k}{1-k}\right).$$

(b) We substitute $k = 0.85$ into the formula from part (a):

$$\text{Total amount spent } = N\left(\frac{k}{1-k}\right) = N\left(\frac{0.85}{1-0.85}\right) = 5.667N.$$

The total additional spending is more than 5 times the size of the original tax rebate.

13. Let k be the fraction of the \$1 bills in circulation that are removed each day. Then $r = 1-k$ is the fraction of bills remaining in circulation the next day. On any day, there are 18 million new bills, and $18r$ million remaining from the day before, and $18r^2$ million remaining from the day before, and so on. In millions, the total number of \$1 bills in circulation is given by

$$N = 18 + 18r + 18r^2 + \cdots.$$

This is an infinite geometric series with $a = 18$ and $-1 < r < 1$ (since the fraction of bills remaining is positive and less than 1). The series converges, and its sum is given by

$$N = \frac{a}{1-r} = \frac{18}{1-r}.$$

Since we know the total number of \$1 bills in circulation is 4 billion, or 4000 million, we have

$$4000 = \frac{18}{1-r}.$$

Solving for r gives

$$1 - r = \frac{18}{4000}$$
$$r = 1 - \frac{18}{4000} = 0.9955.$$

Thus 99.55% of the notes remain in circulation the next day, so 0.45% are removed each day.

Notice that if 18 million new bills are introduced ech day, then at steady state, the same number are removed each day. Thus, we can calculate r without using a series:

$$r = \frac{18 \text{ million}}{4 \text{ billion}} = 0.45\%.$$

Solutions for Section 11.3

1. After receiving the nth injection, the quantity in the body is 50 mg from the nth injection, $50(0.60)$ from the injection the previous day, $50(0.60)^2$ from the injection two days before, and so on. The quantity remaining from the first injection (which has been in the body for $n-1$ days) is $50(0.60)^{n-1}$. We have

$$\text{Quantity after } n^{\text{th}} \text{ injection } = 50 + 50(0.60) + 50(0.60)^2 + \cdots + 50(0.60)^{n-1}.$$

(a) The quantity of drug in the body after the 3rd injection is

$$\text{Quantity after } 3^{\text{rd}} \text{ injection} = 50 + 50(0.60) + 50(0.60)^2 = 98 \text{ mg}.$$

Alternately, we could find the sum using the formula for a finite geometric series with $a = 50$, $r = 0.60$, and $n = 3$:

$$\text{Quantity after } 3^{\text{rd}} \text{ injection } = \frac{50(1-(0.60)^3)}{1-0.60} = 98 \text{ mg}.$$

(b) Similarly, we have

$$\text{Quantity after } 7^{\text{th}} \text{ injection} = 50 + 50(0.60) + 50(0.60)^2 + \cdots + 50(0.60)^6.$$

We use the formula for the sum of a finite geometric series with $a = 50$, $r = 0.60$, and $n = 7$:

$$\text{Quantity after } 7^{\text{th}} \text{ injection} = \frac{50(1-(0.60)^7)}{1-0.60} = 121.5 \text{ mg}.$$

(c) The steady state level is the long term amount of drug in the body if injections are continued indefinitely. Right after an injection is given, we have

$$\text{Steady state level } = 50 + 50(0.60) + 50(0.60)^2 + \cdots.$$

This is an infinite geometric series with $a = 50$ and $r = 0.60$. Since $-1 < r < 1$, the series converges. Its sum is

$$\text{Steady state level } = \frac{50}{1-0.60} = 125 \text{ mg}.$$

5. (a) After a single dose of 50 mg of fluoxetine, the quantity, Q, in the body decays exponentially, so $Q = 50b^t$. After 3 days, half the original dose remains. We use this information to solve for b:

$$\begin{aligned} 25 &= 50b^3 \\ 0.5 &= b^3 \\ b &= (0.5)^{1/3} = 0.7937. \end{aligned}$$

The fraction of a dose remaining after one day is 0.7937.

(b) The level of fluoxetine remaining right after taking the 7th dose is the sum of a finite geometric series with $a = 50$, $r = 0.7937$, and $n = 7$.

$$\begin{aligned} \text{Amount after } 7^{\text{th}} \text{ dose } &= 50 + 50(0.7937) + 50(0.7937)^2 + \cdots + 50(0.7937)^6 \\ &= \frac{50(1-(0.7937)^7)}{1-0.7937} = 194.27 \text{ mg}. \end{aligned}$$

(c) The steady state level of fluoxetine in the body right after taking a dose is the sum of an infinite geometric series with $a = 50$ and $r = 0.7937$. Since $-1 < r < 1$, the series converges. Its sum is

$$\begin{aligned} \text{Steady state level after dose } &= 50 + 50(0.7937) + 50(0.7937)^2 + \cdots. \\ &= \frac{50}{1-0.7937} = 242.37 \text{ mg}. \end{aligned}$$

9. Since the toxin decays at a continuous rate of 0.5% per day, the amount remaining one day after consuming a single amount of 8 micrograms is $8e^{-0.005}$. The person is consuming 8 micrograms every day, so the total accumulated toxin the person has in the body right after consuming the toxin is the sum of 8 (from the amount just consumed) + $8(e^{-0.005})$ (from the amount consumed the previous day) + $8(e^{-0.005})^2$ (from the amount consumed two days ago), and so on. The total accumulation in the body is the sum of an infinite geometric series with $a = 8$ and $r = e^{-0.005}$. Since $-1 < r = 0.9950124 < 1$, the series converges to the sum:

$$\begin{aligned}\text{Total accumulation right after lunch} &= 8 + 8(e^{-0.005}) + 8(e^{-0.005})^2 + \cdots \\ &= \frac{8}{1 - e^{-0.005}} = 1604.0 \text{ micrograms.}\end{aligned}$$

Since the person consumes 8 micrograms each day, the total accumulation of the toxin right before lunch is $1604 - 8 = 1596$ micrograms.

13. If Q_n denotes the total amount of the mineral mined and used in the first n years of the millennium (that is, 2001, 2002, ...), we have

$$Q_n = 5000 + 5000(1.08) + 5000(1.08)^2 + \cdots + 5000(1.08)^{n-1}.$$

This is a finite geometric series with $a = 5000$ and $r = 1.08$, so we have

$$Q_n = \frac{a(1 - r^n)}{1 - r} = \frac{5000(1 - (1.08)^n)}{1 - 1.08} = 62500((1.08)^n - 1).$$

We want the value of n for which the total consumption, Q_n, reaches 350,000 m^3, the total amount available. We can estimate n numerically (using trial and error) or graphically, or we can find n analytically:

$$\begin{aligned}62{,}500((1.08)^n - 1) &= 350{,}000 \\ (1.08)^n - 1 &= 5.6 \\ (1.08)^n &= 6.6.\end{aligned}$$

Taking logarithms and using $\ln(A^p) = p \ln A$, we have

$$\begin{aligned}n \ln(1.08) &= \ln(6.6) \\ n &= \frac{\ln(6.6)}{\ln(1.08)} = 24.5 \text{ years.}\end{aligned}$$

Thus, if consumption of this mineral continues to increase at 8% a year, all reserves will be exhausted in 25 years. If, however, the relative rate of increase changes, the length of time until the reserve runs out may be very different.

17. If consumption changes by a factor of r each year (so we use $r = 0.995$ in part (a) and $r = 1.025$ in part (b)), then consumption in 2001 is predicted to be $27r$ billion barrels, in 2002 it is predicted to be $27r^2$ billion barrels, in 2003 it is predicted to be $27r^3$ billion barrels, and so on. Thus, if Q_n is total consumption in n years, in billions of barrels,

$$Q_n = 27r + 27r^2 + 27r^3 + \cdot + 27r^n.$$

Using the formula for the sum of a finite geometric series, we have

$$Q_n = 27r\frac{1 - r^n}{1 - r}.$$

Since the total reserves are 1000 billion barrels, in each case we find n making $Q_n = 1000$.

(a) Using $r = 0.995$, we solve

$$\begin{aligned}27(0.995)\frac{1 - (0.995)^n}{1 - 0.995} &= 1000 \\ 5373(1 - (0.995)^n) &= 1000 \\ 1 - (0.995)^n &= \frac{1000}{5373} = 0.1861 \\ (0.995)^n &= 0.8139.\end{aligned}$$

Taking logs and using $\ln(A^p) = p \ln A$,

$$\begin{aligned}n \ln(0.995) &= \ln(0.8139) \\ n &= \frac{\ln(0.8139)}{\ln(0.995)} = 41.1 \text{ years.}\end{aligned}$$

(b) Using $r = 1.025$, we have

$$\begin{aligned}
27(1.025)\frac{1-(1.025)^n}{1-1.025} &= 1000 \\
1107((1.025)^n - 1) &= 1000 \\
(1.025)^n - 1 &= \frac{1000}{1107} = 0.9033 \\
(1.025)^n &= 1.9033 \\
n\ln(1.025) &= \ln(1.9033) \\
n &= \frac{\ln(1.9033)}{\ln(1.025)} = 26.1 \text{ years.}
\end{aligned}$$

Solutions for Chapter 11 Review

1. The formula for the sum of a finite geometric series with $a = 100$, $r = 0.85$, and $n = 11$ gives

$$\text{Sum} = 100 + 100(0.85) + \cdots + 100(0.85)^{10} = \frac{100(1-(0.85)^{11})}{1-0.85} = 555.10.$$

5. Since $6300/31500 = 0.2$ and $1260/6300 = 0.2$ and $252/1260 = 0.2$, this is an infinite geometric series with $a = 31500$ and $r = 0.2$:

$$\text{Sum} = 31500 + 6300 + 1260 + 252 + \cdots = 31500 + 31500(0.2) + 31500(0.2)^2 + 31500(0.2)^3 + \cdots.$$

Since $-1 < r < 1$, the series converges and the sum is given by:

$$\text{Sum} = \frac{31500}{1-0.2} = 39{,}375.$$

9. **(a)** The amount of atenolol in the blood is given by $Q(t) = Q_0 e^{-kt}$, where $Q_0 = Q(0)$ and k is a constant. Since the half-life is 6.3 hours,

$$\frac{1}{2} = e^{-6.3k}, \quad k = -\frac{1}{6.3}\ln\frac{1}{2} \approx 0.11.$$

After 24 hours

$$Q = Q_0 e^{-k(24)} \approx Q_0 e^{-0.11(24)} \approx Q_0(0.07).$$

Thus, the percentage of the atenolol that remains after 24 hours $\approx 7\%$.

(b)

$$\begin{aligned}
Q_0 &= 50 \\
Q_1 &= 50 + 50(0.07) \\
Q_2 &= 50 + 50(0.07) + 50(0.07)^2 \\
Q_3 &= 50 + 50(0.07) + 50(0.07)^2 + 50(0.07)^3 \\
&\vdots \\
Q_n &= 50 + 50(0.07) + 50(0.07)^2 + \cdots + 50(0.07)^n = \frac{50(1-(0.07)^{n+1})}{1-0.07}
\end{aligned}$$

(c)

$$
\begin{aligned}
P_1 &= 50(0.07)\\
P_2 &= 50(0.07) + 50(0.07)^2\\
P_3 &= 50(0.07) + 50(0.07)^2 + 50(0.07)^3\\
P_4 &= 50(0.07) + 50(0.07)^2 + 50(0.07)^3 + 50(0.07)^4\\
&\vdots\\
P_n &= 50(0.07) + 50(0.07)^2 + 50(0.07)^3 + \cdots + 50(0.07)^n\\
&= 50(0.07)\left(1 + (0.07) + (0.07)^2 + \cdots + (0.07)^{n-1}\right) = \frac{0.07(50)(1-(0.07)^n)}{1-0.07}
\end{aligned}
$$

13. (a) (i) On the night of December 31, 1999:

First deposit will have grown to $2(1.04)^7$ million dollars.
Second deposit will have grown to $2(1.04)^6$ million dollars.
...
Most recent deposit (Jan.1, 1999) will have grown to $2(1.04)$ million dollars.

Thus

$$
\begin{aligned}
\text{Total amount} &= 2(1.04)^7 + 2(1.04)^6 + \cdots + 2(1.04)\\
&= 2(1.04)\underbrace{(1 + 1.04 + \cdots + (1.04)^6)}_{\text{finite geometric series}}\\
&= 2(1.04)\left(\frac{1-(1.04)^7}{1-1.04}\right)\\
&= 16.43 \text{ million dollars.}
\end{aligned}
$$

(ii) Notice that if 10 payments are made, there are 9 years between the first and the last. On the day of the last payment:

First deposit will have grown to $2(1.04)^9$ million dollars.
Second deposit will have grown to $2(1.04)^8$ million dollars.
...
Last deposit will be 2 million dollars.

Therefore

$$
\begin{aligned}
\text{Total amount} &= 2(1.04)^9 + 2(1.04)^8 + \cdots + 2\\
&= 2\underbrace{(1 + 1.04 + (1.04)^2 + \cdots + (1.04)^9)}_{\text{finite geometric series}}\\
&= 2\left(\frac{1-(1.04)^{10}}{1-1.04}\right)\\
&= 24.01 \text{ million dollars.}
\end{aligned}
$$

(b) In part (a) (ii) we found the future value of the contract 9 years in the future. Thus

$$\text{Present Value} = \frac{24.01}{(1.04)^9} = 16.87 \text{ million dollars.}$$

Alternatively, we can calculate the present value of each of the payments separately:

$$
\begin{aligned}
\text{Present Value} &= 2 + \frac{2}{1.04} + \frac{2}{(1.04)^2} + \cdots + \frac{2}{(1.04)^9}\\
&= 2\left(\frac{1-(1/1.04)^{10}}{1-1/1.04}\right) = 16.87 \text{ million dollars.}
\end{aligned}
$$

Notice that the present value of the contract ($16.87 million) is considerably less than the face value of the contract, $20 million.

17.

$$\text{Present value of first coupon} = \frac{50}{1.04}$$
$$\text{Present value of second coupon} = \frac{50}{(1.04)^2}, \text{etc.}$$

$$\begin{aligned}
\text{Total present value} &= \underbrace{\frac{50}{1.04} + \frac{50}{(1.04)^2} + \cdots + \frac{50}{(1.04)^{10}}}_{\text{coupons}} + \underbrace{\frac{1000}{(1.04)^{10}}}_{\text{principal}} \\
&= \frac{50}{1.04}\left(1 + \frac{1}{1.04} + \cdots + \frac{1}{(1.04)^9}\right) + \frac{1000}{(1.04)^{10}} \\
&= \frac{50}{1.04}\left(\frac{1 - \left(\frac{1}{1.04}\right)^{10}}{1 - \frac{1}{1.04}}\right) + \frac{1000}{(1.04)^{10}} \\
&= 405.545 + 675.564 \\
&= \$1081.11
\end{aligned}$$

NOTES

NOTES

NOTES

NOTES

NOTES

NOTES

NOTES

NOTES

NOTES

NOTES

NOTES